INTERNATIONAL SERIES OF MONOGRAPHS IN
PURE AND APPLIED BIOLOGY

Division: **ZOOLOGY**

GENERAL EDITOR: G. KERKUT

VOLUME 1

AN OUTLINE OF
DEVELOPMENTAL PHYSIOLOGY

AN OUTLINE OF
DEVELOPMENTAL PHYSIOLOGY

by

 CHR. P. RAVEN

PROFESSOR OF ZOOLOGY IN THE UNIVERSITY
OF UTRECHT

Translated by
L. DE RUITER, biol. drs.

PERGAMON PRESS

OXFORD · LONDON · EDINBURGH · NEW YORK
TORONTO · PARIS · BRAUNSCHWEIG

Pergamon Press Ltd., Headington Hill Hall, Oxford
4 & 5 Fitzroy Square, London W.1

Pergamon Press (Scotland) Ltd., 2 & 3 Teviot Place, Edinburgh 1

Pergamon Press Inc., 44-01 21st Street, Long Island City, New York 11101

Pergamon of Canada, Ltd., 6 Adelaide Street East, Toronto, Ontario

Pergamon Press S.A.R.L., 24 rue des Écoles, Paris 5ᵉ

Vieweg & Sohn GmbH, Burgplatz 1, Braunschweig

First published in the Dutch language 1948
First English edition 1954
Second (revised) English edition 1959
Third impression 1961
Third (revised) English edition 1966

Printed in the Netherlands by J. Noorduijn en Zoon N.V.

(2364/66)

Contents

Preface

This book was first published in the Dutch language in 1948, while the first English edition appeared in 1954. It was meant in the first place for readers who, though interested in its subject and having some general knowledge of science, were not acquainted with more than the first elements of biology. Therefore, highly technical digressions were avoided, and it was attempted to construct, as far as possible, a well-rounded picture of the phenomena of development. This made it not always possible to make a sharp distinction between well-founded facts and more or less hypothetical trains of thought which link them together. This would have required a much more exhaustive and critical treatment than could be given here. Therefore, the picture of development drawn here is a more or less subjective one.

It follows that the book, according to its design, has never been meant as a textbook. On the other hand, it may be used as a preliminary introduction into the field in undergraduate classes, and in fact it has been frequently used as such. Therefore, a list of references has been added in the interest of those who might want to refer to the original publications on some special problem. Moreover, a glossary of scientific terms has been included for the benefit of those readers who are not professional biologists.

This new edition has been brought up to date by taking into account the literature up to about the middle of 1963. In the list of references several important papers, which have appeared since the previous edition of the book, have been included; some older papers have been expunged in order not unduly to enlarge the list.

Chr. P. R.

CHAPTER I

Introduction

"Developmental physiology" is the term that will be applied here to the branch of biology that studies the causal relationships in animal development. About 1880, Wilhelm Roux founded this branch of science on the strength of theoretical considerations and a small number of specially designed experiments. He called it "developmental mechanics". This name, however, might well give rise to misconceptions as to its nature. Roux applied the term "mechanics" in the wide sense of "theory of causal connections" but its use may easily lead to the idea that an attempt is being made to explain the developmental phenomena on the basis of the laws of mechanics. For that reason we prefer the name "developmental physiology", which has the same scope. This means that the word "physiology" is used here in its widest sense, viz. that of "causal science of living organisms", so that it does not imply that we shall study only those processes during development that are normally included in "physiology" in its narrower sense, such as metabolism, excitability, etc. The term "development" is meant to include all irreversible changes that the organism goes through in the course of its existence, from the moment of its origin until the death of the individual.

In broad outline, the development of all multicellular animals proceeds along the same route. We shall give a short account of this now, taking the *fertilised egg* as the starting point of development. This arises from the fusion of a male and a female germ cell (called *sperm* and *egg* respectively). (For the present we shall disregard special cases such as that of vegetative reproduction, in which the new individual originates from a cell or cell group that separates itself from the body of an animal.) The first developmental process is a division of the fertilised

1

egg into cells. This so-called *cleavage* results in the formation of a small lump of more or less uniform cells, the *blastomeres*. A cavity develops in the centre of this cell lump, and the vesicular germ is then called a *blastula*. In the period following cleavage the germ goes through important changes, consisting of a series of movements of cell groups. The blastomeres are thereby arranged in several layers which in their turn will divide into more or less well defined masses of cells. Together these processes may be called the *topogenesis* of the embryo; they serve to divide the germ into a number of *organ primordia,* each of which contains the material for one definite organ or group of organs of the embryo. At first, these primordia consist of cells which are all very similar, and of an almost undifferentiated, embryonic character. This changes during the next phase of development, when each cell of the primordia specialises in a particular direction; in this way the various tissues of the body are formed. This is called the period of tissue differentiation (*histogenesis*). Once development has proceeded so far, the organs and tissues of the embryo begin to take up their respective functions. Development has by no means ended yet, but in the subsequent period the functions of the organs play a decisive role in their differentiation. This is therefore called the *functional stage* of development. Finally, two categories of change which occur at a very late stage may be mentioned, viz. those which lead to full *maturity* of the organism, and those of *senility*. The latter are mainly of a disintegrative character, and eventually result in the death of the individual. A further category that must be specially mentioned is formed by developmental processes that do not manifest themselves during the normal, undisturbed life of the individual, but may lead to a more or less complete regeneration, a *restitution* of the structure of the organism when this has been damaged.

When we survey the outline of the course of development just given, its most striking characteristic appears to be an *increase in spatial multiplicity*. This is seen most clearly if we compare the culminating point of development, the adult individual, with its starting point, the fertilised egg. On the one hand we have a very complicated whole of organs, tissues and cells. Each of these has its own place in the organism, and its own characteristic, orderly structure. Together they form an integrated system of a very high degree of multiplicity. On the other hand we see a small lump of protoplasm with a nucleus, very simple in shape, and practically without a visible structure. Clearly the "spatial multiplicity" of the organism has increased greatly during its development. Yet this increase might be apparent only, for the structure of the egg, though invisible and therefore seemingly simple, could in reality be very complicated. In past centuries, it has indeed been assumed that the egg contained the future organism in its full complexity, although not in an easily visible form. This is the theory of *Preformation* (*Evolutio*). Later this concept was modified as follows. It was assumed that, though the future animal as such is not present in the egg, still each part of the animal is represented by a corresponding part of the egg, which contains the factors necessary for its formation, while inversely each part of the egg can only give rise to one part of the animal. This implies that there is a one-to-one relation between the parts of the egg and adult, and both have the same degree of ordered multiplicity (W. Roux's theory of *Neo-Evolutio*). We shall see in Chapter III that experimental investigations have shown that this view cannot be maintained. It has been demonstrated that the structure of the egg is really very simple, and that it contains no counterpart of the complicated structure of the adult animal. The latter is only gradually built up during development. There is therefore a real increase in ordered spatial multiplicity during development.

This statement needs some qualification, however. In the first place, the word *spatial* must be stressed, for there can be no doubt that the protoplasm of the egg is a highly complicated mixture of substances. From that point of view it has a great multiplicity already. This, however, is not a spatial multiplicity for the various components are not restricted to fixed places within the system but are distributed more or less homogeneously over the egg instead. We will call this *non-spatial* or *intensive* multiplicity, as opposed to *spatial* or *extensive* multiplicity in which the various parts are arranged side by side in space.[1] To a certain extent, therefore, development may be characterised as a transition from intensive to extensive multiplicity.

Secondly, it must be taken into consideration that a certain amount of spatial multiplicity is contained in the structure of the large organic macromolecules that form part of the substance of the egg cell. The possibility must be taken into account that this plays a part in development, which, in that case, might partly represent a transformation or translation of this invisible, molecular structure of the egg cell into macroscopic structure of the adult animal.

We will briefly indicate here how we can imagine such a development, in which intensive multiplicity changes into extensive multiplicity, or molecular structure is translated into macroscopic structure, to take place. This will at the same time illustrate the line of thought underlying this book. Before we start, however, two points must be made clear. First, it must be pointed out that development follows very different lines in different animal groups. It is only in broad outline that all these different modes of development are governed by the same laws. Only these general features can be sketched here, and naturally we shall often have to resort to broad generalisation in so doing. In the second place, it must be remembered that developmental physiology is a young branch of biology, and that it has by no means yet constructed a generally accepted and balanced system of ideas. Therefore, the best attempt at a well rounded picture of development we can now make will still be of a very subjective and preliminary nature, and will be open to many future changes and improvements. The need for generalisation entails the danger of partiality. Moreover, it will be necessary now and then to draw conclusions from factual data that are at present insufficient, and to bridge gaps in our knowledge by means of hypothetical constructions. We can base such opinions only upon the facts that have so far become known to science. Therefore, the reader must take care

---

[1] H. Driesch has used these terms in another sense. He attributes a metaphysical meaning to the concept of "intensive multiplicity". It should be stressed that this is not implied in the present use of the word.

not to regard the picture of animal development given in this book as definitive and irrefutable. We have only tried to place the results of the several experiments in such a context that a wider circle of readers may appreciate the importance of the phenomena discovered.

The fusion of sperm and egg starts a number of processes which inaugurate development (Ch. II). The structure of the fertilised egg is still very simple; the various components of the cytoplasm are more or less evenly distributed so that all parts of the egg are still approximately equivalent (Ch. III). Yet the main axes of the egg are already fixed; it has a polarity, and often a bilateral symmetry. This polarity and symmetry are probably localised partly in the more solid outer layer, or *cortex* of the egg (Ch. IV). Next, certain components of the egg cytoplasm (so-called *determining substances*) begin to accumulate locally under the influence of these cortical factors. This leads to the onset of heterogeneity in the egg, for now its parts begin to vary in chemical composition. This process is called the *chemodifferentiation* of the egg (Ch. V). The variation in chemical composition within the egg influences other properties of the protoplasm as well, such as permeability, metabolism, etc. Moreover, the determining substances begin to interact with one another. In this way the spatial multiplicity of the egg-system increases rapidly. Now the *genes,* the carriers of the hereditary properties which are localised in the nucleus, begin to intervene. In the course of cleavage, the zygote nucleus which arose from the fusion of the nuclei of egg and sperm has divided into a great number of segmentation nuclei. These come to be situated in parts of the cytoplasm which vary in physicochemical composition as a consequence of chemodifferentiation. By an interaction with the surrounding cytoplasm certain genes are "activated" in each segmentation nucleus, whereas others remain inactive. The activated genes play an important part in synthetic processes, especially of proteins. In different cells different proteins, often having the properties of enzymes, are synthesised. These may in their turn affect the course of various other metabolic processes in the cells. Part of the substances produced remain in the cells, and determine their further development. Other substances, however, may diffuse from tne cells, and exert their influence in other parts of the body (Ch. VI). Under the influence of the ever-increasing chemodifferentiation, divergent specialisation takes place in the motility of the protoplasm of the various cell groups. The different parts of the blastula vary in this respect, and consequently a system of shifts of cell groups occurs, which we have already named the *topogenesis* of the embryo (Ch. VII). This in the first place moves the material for the future organ primordia to the appropriate places in the embryo. Secondly, however, it establishes direct contact between cell groups of different physico-chemical composition which were spatially separated at first. These cell groups thereby influence each other (*induction*), and this interaction results in a rapid increase in the chemodifferentiation of the embryo. Here a leading role may sometimes be played by particular cell groups (*organisers*), distinguished perhaps by a higher content of certain determining substances (Ch. VIII). In the following period, a complicated system of topogenetic and inductive processes transforms the embryo into a mosaic of organ primordia. Each of these has different physicochemical properties, and consequently different developmental potencies. This stage is succeeded by the period of *tissue differentiation,* during which the cells of the primordia develop into definite cell- and tissue structures, as determined by their previous chemodifferentiation (Ch. IX). Later the structure of organs and tissues may be further perfected under the influence of their functions (*functional adaptation*) (Ch. X). Finally, some sort of an equilibrium is achieved. From then on, developmental processes continue to occur in restricted regions of the body only. In many animals, however, a disturbance of this equilibrium by the removal of part of the body leads to renewed developmental processes which result in a more or less complete *regeneration* of the lost part (Ch. XI).

CHAPTER II

The initiation of development:
The fertilisation of the egg

In sexual reproduction, which occurs in all multicellular animals, two types of germ cells or *gametes* are produced by the adult individuals. In some cases both types are produced by the same individual, in other cases they are formed by different animals. The two types are female germ cells, or *eggs*, and male germ cells, or *sperms* (spermatozoa). Each egg fuses with a sperm; this is the process of *fertilisation*. The fertilised egg, or *zygote*, develops into a new individual.

As a rule, the egg is a very large cell, consisting of protoplasm with a large nucleus, commonly called *germinal vesicle*. In the protoplasm, a great quantity of food, in the form of protein and fat globules, is often accumulated. This is the so-called *yolk*, which constitutes a store, given to the young individual as it sets out upon its path of life. It supplies the energy for the first stages of development, during which the embryo cannot yet take up food from its environment. The eggs originate from small cells, the so-called *oögonia*, in the *ovary* of the maternal animal. They increase strongly in size in the course of the accumulation of the yolk substances, and grow into immature eggs, or *oöcytes*.

In contrast with the eggs, the sperms are very small, usually thousands of times smaller than the egg; they do not contain any yolk. Apart from slight changes in shape, eggs as a rule are immobile; sperms, on the other hand, have the power of locomotion. This enables them to move actively towards the eggs, and to penetrate into them, which results in fertilisation. The shape of the sperms varies rather considerably among the different groups of animals, but usually we can distinguish a head, representing the nucleus of the cell, a middle piece, and a long, motile tail. The undulating movements of the latter propel the sperm.

In all organisms, the nucleus of each cell contains a number of small bodies with a high affinity for stains, the *chromosomes*. They become visible each time the nucleus divides. Within each species, their number is constant. We shall see later that they are very important for the life of the organism (Ch. VI). During fertilisation the nuclei of egg and sperm fuse (*amphimixis*), so that a double number of chromosomes is present in the nucleus of the fertilised egg, and likewise in that of all the cells of the new individual, which arise from it by cell division. This would lead to doubling of the chromosome number in each successive sexually reproduced generation, but for the fact that during the formation of the germ cells the number is each time reduced again to the single value. This takes place in the so-called *reduction division*, actually two divisions in quick succession in the course of which the double (*diploid*) set of chromosomes is reduced to a single (*haploid*) set. In the formation of sperms, these reduction divisions take place in the male gonad, or *testis*, one diploid *spermatocyte* giving rise to four haploid *spermatids* which then change into spermatozoa. In eggs, the reduction process takes a slightly different course. It is known as *maturation*, and proceeds as follows. The nucleus of the oöcyte moves towards the surface; its nuclear membrane disappears, and a mitotic spindle develops. Meanwhile each chromosome has duplicated itself, and the double chromosomes now arrange themselves in pairs (*tetrads*) in the centre of the spindle (Fig. 1a). The two halves of each tetrad then move apart, so that two equal groups of double chromosomes are formed. One of these groups, together with a small quantity of protoplasm, is expelled from the egg, and forms the *first polar body* (Fig. 1b). Immediately afterwards, a new spindle forms in the egg, and the remaining double chromosomes (*dyads*) arrange themselves in the centre of this spindle. The two halves of each dyad separate (Fig. 1c). One group is expelled again, forming the *second polar body*. From the remaining chromosomes a new nucleus is formed, the so-called *female pronucleus* (Fig. 1d).

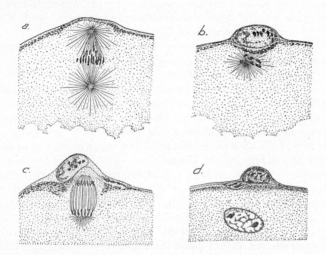

Fig. 1. Maturation divisions in the egg of a starfish. (*a*) first maturation division; (*b*) extrusion of the first polar body; (*c*) second maturation division; (*d*) mature egg (below: pronucleus, at top: first and second polar body). After Buchner.

This maturation of the egg may occur at different times. Sometimes both maturation divisions take place within the ovary or in the genital ducts of the maternal animal. In other cases, maturation begins in the ovary, but comes to a stop at a certain stage, and is not completed until after fertilisation. Finally, in many species the female lays unripe oöcytes, and the whole process of maturation takes place outside the maternal body. In a number of instances of this last type, the nature of the stimulus which causes the onset of maturation has been investigated. Sometimes the maturation divisions do not begin until after the sperm has penetrated into the egg. In other cases, maturation occurs a few minutes after the oöcytes have been laid in sea water. It has been found that the calcium ions present in sea water play an important role (Dalcq, Pasteels, Heilbrunn). These ions, possibly in combination with potassium ions, appear to provide the specific stimulus which causes the membrane of the germinal vesicle to disappear, and which thereby initiates maturation.

In normal development, the polar bodies are always formed at one pole of the egg, the so-called *animal pole* (p. 43). At the beginning of maturation, the germinal vesicle, or the maturation spindle arising from it, moves towards this pole; and the spindle there orients itself at right angles to the egg surface (Fig. 1a). Apparently this part of the surface attracts the spindle. In several cases it has been possible to weaken this attraction by a special treatment of the egg, e.g. by means of an excess of calcium ions in the medium in starfishes and in some molluscs, or by means of a rise of temperature in sea urchins. Dalcq (1924) has called this *depolarisation*. As a result the maturation spindle does not come close enough to the egg surface, and consequently abnormally large polar bodies are formed. In the case of an even stronger depolarisation the maturation spindle does not orient itself at right angles to the surface, and the polar bodies fail to appear altogether.

The maturation of the egg has still another consequence. Once the nuclear membrane has dissolved, the nucleoplasm contained in the germinal vesicle mixes with the cytoplasm. This causes a change in several of the physical and physiological properties of the cytoplasm (e.g. its viscosity and permeability). Moreover, it appears that many eggs cannot be fertilised until this has happened. No sperms can penetrate into immature oöcytes, or, if they can, they remain inactive in the egg cytoplasm for the time being, and do not "awake" until maturation has begun. Costello (1940), working with fragmented oöcytes of the polychaete worm *Nereis*, showed that only those parts that contained the germinal vesicle could be successfully fertilised. In other animals, e.g. starfishes and sea urchins, it is equally impossible to fertilise non-nucleated egg fragments, if the germinal vesicle was intact at the moment of fragmentation. However, if maturation had already set in at that time, non-nucleated fragments can be successfully fertilised, and will

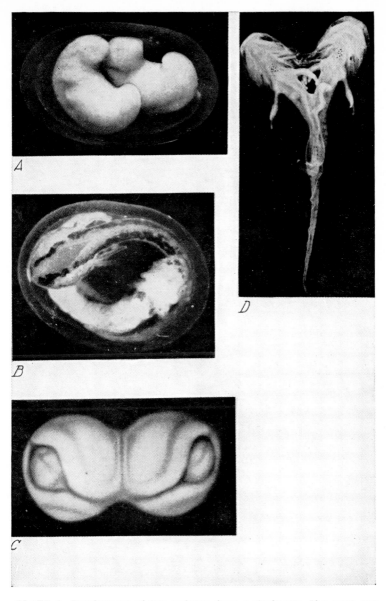

PLATE I. Development of two embryos from a single egg. If a newt egg is divided into two halves by a ligature (cf. Fig. 13), both halves may under certain circumstances develop into embryos (**A-B**). In the case of incomplete constriction, the resulting twins form a double monster (**C-D**). After Spemann.

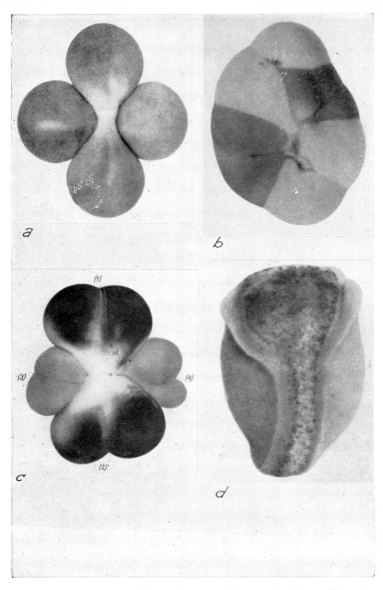

PLATE II. Development of a single embryo from two fused eggs. Crosswise fusion (a) of two newt eggs at the 2-cell stage results in the production of a single embryo (b), the four quadrants of which arise from one or the other of the original eggs alternately (c). Even eggs of different species (**Triton taeniatus** and **T. alpestris**), fused in this way, may produce a single embryo (d). After Mangold and Seidel.

go on developing (so-called *merogony*). This was found by Delage as early as 1899. All this shows that the mixing of nuclear sap and cytoplasm at the beginning of maturation is of great importance for fertilisation.

In the fusion of egg and sperm, substances secreted by each of these cells play a major role. Many eggs (e.g. of echinoderms, molluscs, annelids, tunicates and vertebrates) have been shown to secrete a substance, generally called *fertilizin,* which accelerates the movements of the sperms, and in high concentrations causes them to "agglutinate". It has been shown that the fertilizin of the sea urchin egg is identical with material of the gelatinous coat surrounding the egg. It is an acid mucopolysaccharide, in which the carbohydrate component differs among various species of sea urchins. The sperms contain a substance, called *antifertilizin,* which neutralises the agglutinating action of fertilizin, and causes a precipitation and contraction of the egg jelly. Antifertilizin is an acidic protein, which seems to be located on the lateral surface of the sperm head. It is assumed that the reaction between fertilizin and antifertilizin is concerned in the initial attachment of the spermatozoon to the egg. Investigations by Tyler (1948) have demonstrated that this interaction has a great similarity to immunological reactions.

Once the sperm has become attached to the egg, it must be able to penetrate the envelopes that surround the latter. It has been shown that the sperms of various species of vertebrates and invertebrates contain lytic agents possessing the property of breaking down the egg membranes. It appears that these *egg-membrane lysins* are often located in the anterior part of the sperm head, the so-called *acrosome.* On making contact with the surface of the egg membranes, in many species the acrosome "explodes", so to speak, thereby giving rise to a straight, rigid filament, which pierces the egg envelopes and reaches the surface of the egg cell proper (Colwin & Colwin, 1957). At the same time, the egg-membrane lysin is released, which erodes the membrane barrier in the vicinity of the sperm head. The egg is stimulated by the acrosome filament making contact with its surface. In many species, it forms a

surface projection, the so-called *fertilisation cone,* which surrounds the acrosome filament like a sleeve, and engulfs the
sperm (Figs. 2, 3).

Colwin and Colwin (1961) have made a very detailed study
by means of electron microscopy of the penetration of the spermatozoon into the egg of the annelid worm *Hydroides.*

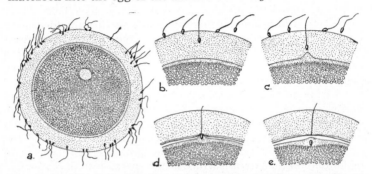

Fig. 2. Fertilisation of the egg of a starfish. (*a*) the egg in its envelopes, surrounded by spermatozoa; (*b-e*) a sperm pierces the
egg envelope and penetrates into the "fertilisation cone". The initial
stages of the formation of a perivitelline cavity between egg and
vitelline membrane. After Fol and Wilson.

The situation in mammals deserves special mention. When
leaving the ovary, the eggs of this group are surrounded by a
layer of follicle cells. These cells are kept together by a viscous
substance, probably consisting mainly of hyaluronic acid, or
a related polysaccharide. Now it has been proved that the
sperms contain an enzyme, *hyaluronidase,* which breaks down
hyaluronic acid, and thereby causes the disintegration of the
cell layer so that the sperms gain free access to the egg surface.

The penetration of the sperm causes the egg to "awake" at
once. Within a few seconds, it reacts with an instantaneous
change of properties. Usually the formation of a *fertilisation
membrane* is the first visible effect. The unfertilised egg is
as a rule surrounded by a *vitelline membrane,* which is sometimes fairly thick, sometimes however barely visible. Sooner or
later after the penetration of the sperm, this membrane begins

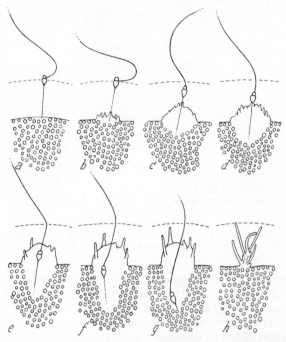

Fig. 3. Successive stages of sperm entry in the egg of a holothurian. (*a*) acrosome filament has pierced the egg jelly and reached the egg surface; (*b-g*) formation of fertilisation cone around the acrosome filament. The sperm moves inward and is engulfed by the fertilisation cone; (*h*) withdrawal of fertilisation cone after sperm entry. After Colwin and Colwin.

to lift from the egg surface (Figs. 2d-e, 4), because substances expelled by the egg accumulate in the space between egg surface and membrane, the so-called *perivitelline space*. These substances, part of which are colloids, take up water from the environment by osmosis. Consequently, the perivitelline space increases rapidly. The extrusion of these substances often results in a considerable decrease in size of the egg.

Usually the formation of the fertilisation membrane does not

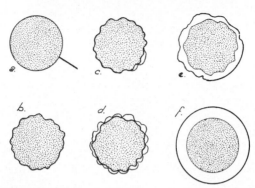

Fig. 4. Formation of the fertilisation membrane in the sea urchin **Strongylocentrotus purpuratus** (the jelly layer around the egg has been omitted). (a) sperm in contact with the egg, about 2 seconds after insemination; (b-e) 30-50 seconds after insemination, formation and coalescence of blebs at the egg surface; (f) 2-3 minutes after insemination, the fertilisation membrane has been formed. After Chase.

set in simultaneously in all parts of the egg surface, but starts from the spot where the sperm has penetrated (Fig. 4c), and proceeds from there in all directions over the egg. Obviously, the extrusion of the perivitelline substances, which causes the elevation of the fertilisation membrane, must be due to a sudden change in the outer layer of the egg, its so-called *cortex*. In sea urchins, this cortical reaction has been studied in great detail and by various methods. The first visible effect is a change in the optical properties (light scattering, birefringence) of the cortex. Starting after a few seconds from the point where the fertilising sperm has attached to the egg surface, it spreads over the egg surface in a wave-like progression, reaching the opposite pole within about twenty seconds. Immediately afterwards, the formation of the fertilisation membrane begins at the point of sperm penetration, and propagates in the same way, and at the same rate, around the egg. The unfertilised sea urchin egg is not surrounded by a distinct vitelline membrane. The fertilisation membrane arises here by splitting of the outer plasma membrane of the egg in two lamellae. Immediately afterwards, particular cortical granules, situated in one layer in vacuoles beneath the plasma membrane, extrude their contents into the perivitelline space between the lamellae, and coalesce with the inside of the outer lamella. The latter becomes the fertilisation membrane, while the inner lamella, together with the walls of the vacuoles of the cortical granules, forms the definitive egg surface. Chemically, the cortical granules consist largely of mucopolysaccharides. They react with the proteins of the membrane, thereby causing a structural change which probably involves both interlinking by side chains, and stretching of the protein molecules. In this way, the fertilisation membrane undergoes a process of consolidation, after which it resists further stretching. Both the propagation of the cortical reaction and the hardening of the fertilisation membrane are dependent upon the presence of calcium ions.

In fishes and frogs the formation of the fertilisation membrane occurs in a similar way, but in these groups there is a preformed vitelline membrane which is lifted from the egg surface at fertilisation.

In the marine polychaete worm *Nereis*, too, cortical granules are extruded during fertilisation (Costello, 1949). Here, however, they coalesce into a substance, the greater part of which passes through the vitelline membrane, and forms a thick gelatinous mantle on the outside of the latter. This jelly is again a polysaccharide.

In general, there is a great deal of variation in these processes among various groups (cf. Pasteels, 1961).

Apart from bringing about the formation of the fertilisation membrane, the cortical reaction is very important in another respect as well. In the great majority of animals, normally only one sperm finds its way into the egg; penetration of more than one sperm into the egg, so-called *polyspermy*, causes abnormal development. In a few groups only, e.g. in birds, we find physiological polyspermy, i.e. penetration of several sperms into one egg, is normal and does not lead to disturbances in development.

Where monospermy is normal, there must be a mechanism that prevents the penetration of more than one sperm. This mechanism has been studied in detail by Rothschild and Swann in sea urchins. They conclude from their experiments that the block to polyspermy is probably diphasic. At the moment of attachment of the fertilising spermatozoön to the egg, a rapid change in cortical structure sweeps over the egg surface in less than two seconds. This reduces the chance of penetration of a second sperm by a factor of 20, but does not make the egg completely impermeable to spermatozoa. It is followed by a slower process, probably consisting in the production of a sperm-impermeable layer at the egg surface, which is established in about 60 seconds (Rothschild, 1954). Later experiments by others have not confirmed the existence of the first rapid process. Presumably, the visible cortical reaction corresponds to the second phase of Rothschild and Swann. The breakdown of the cortical granules may contribute to the formation of the sperm-impermeable layer. The integrity of this layer depends on the existence of divalent cations in the external medium. If they are not present, the layer dissolves and the eggs can be refertilised. It is probably identical with the so-called *hyaline layer,* which becomes visible on the surface of the sea urchin egg a few seconds after the cortical reaction has passed over it. Apart from its importance for the establishment of the block to polyspermy, the hyaline layer plays an important role during later development as a membrane keeping the cells together during division.

We have seen above (p. 11) that the fertilizins of different species of sea urchins differ in their chemical composition. It has been shown that the action of fertilizin is predominantly species-specific. This may be regarded as a barrier against hybrid fertilisation, but it does not completely prevent it. Closely related species can often be crossed easily. By means of such artifices as high sperm concentrations, and modification of the acidity of the medium, it is possible to combine eggs and sperm of species which are very far apart taxonomically. In this way, sea urchins' eggs have been successfully fertilised, even with sperm of worms or molluscs (Kupelwieser, Godlewski). We shall later return to these experiments.

The sperm penetrates head first into the egg, but soon after-

wards it rotates through 180°, so that its head points towards
the egg surface, and its middle piece towards the centre. In
those cases where the tail of the sperm has entered into the
egg as well, this now drops off and dissolves in the egg cyto-

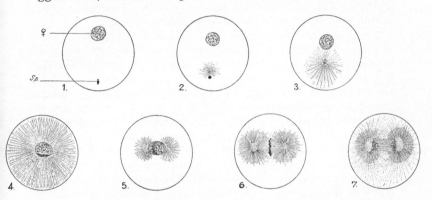

Fig. 5. Cytology of fertilisation in a sea urchin. 1-4: formation of
the sperm aster, fusion of the nuclei of egg and sperm; 5-7: formation of the amphiaster, first cleavage division. ♀ : female pronucleus.
sp: sperm nucleus. After Fry and Parks.

plasm. The head of the sperm swells strongly, and, regaining
the appearance of a normal nucleus, becomes the *male pronucleus*. Originating from the middle piece of the sperm, a
star-shaped radiation, the *sperm aster*, appears in the egg
cytoplasm (Fig. 5, 1-3). Chambers has shown by microdissection
that this is due to local gelation of the cytoplasm. This process
of gelation spreads like a wave in all directions through the
protoplasm. At the same time, the centre of the sperm aster
liquefies again, forming an area of liquid protoplasm around
the male pronucleus. The male and female pronuclei move
towards each other, and meet. They may lie side by side for some
time, or, alternately, they may fuse immediately (Fig. 5, 4).
Now two new radiations appear, one on each side of the pair
of pronuclei (Fig. 5, 5). They meet, forming a so-called *amphiaster*. Then the nuclear membranes of both pronuclei disappear,

thereby liberating their chromosomes. A mitotic spindle develops between the poles of the amphiaster, and the chromosomes arrange themselves in the centre of the spindle (Fig. 5, 6-7), thereby completing the union of the two nuclei (*amphimixis*). The next step will be a cell division, the first *cleavage*, which marks the beginning of the development of the germ.

A number of experiments have provided information as to the factors governing the processes here described. First, it appears that at least two conditions must be fulfilled for the development of the sperm within the egg to take place: (1) the maturation of the egg must have begun, so that the nuclear sap of the germinal vesicle has mixed with the egg cytoplasm (p. 10); (2) the sperm must have traversed the egg cortex. Sperms that have been artificially injected into the egg remain completely inactive (Hiramoto, 1962). The same is true of sperms which have penetrated into egg fragments without cortical plasm (Chambers). Apparently the passage of the sperm through the egg cortex, with the ensuing cortical reaction of the egg, form a prerequisite for the further development of the sperm within the egg.

Allen and Hagström (1955) succeeded in obtaining sea urchin eggs in which the cortical reaction had not propagated over the entire surface of the egg, by interrupting the fertilisation reaction by a heat shock within 20 seconds after sperm attachment. Restriction of the amount of fertilised cortex around the point of sperm penetration limits the growth of the sperm aster and consequently the distance of penetration of the sperm nucleus. The egg nucleus failed to respond to the presence of the sperm if either one of the two pronuclei was largely surrounded by unfertilised cortex. These results give evidence of a change in the physiological conditions of the egg cytoplasm depending on the cortical reaction, which is essential for the normal course of the processes bringing about the fusion of the pronuclei.

The best study of these processes was made by Fankhauser (1925-41) in newts. In this group, there is so-called facultative polyspermy, i.e. more than one sperm may penetrate into one egg without disturbing the development. One of the male pronuclei combines with the nucleus of the egg. The other sperms which have penetrated show some initial development, but then they disintegrate. Fankhauser constricted eggs shortly after fertilisation, by means of a hair ligature, dividing them either completely or partially into two halves. One half contained the egg nucleus, with or without one or more sperm nuclei. The other half contained only sperm nuclei.

Fankhauser observed that, if constriction was complete, a number of sperms developed simultaneously, forming large asters in the half without the egg nucleus. In the case of incomplete constriction, what happened in this half of the now dumb-bell shaped egg depended upon whether or not, in the other half, the egg nucleus had meanwhile united with a sperm nucleus. If this had taken place, development occurred mainly in those sperms in the half without the egg nucleus that were farthest removed from the peduncle which joined the two halves (Fig. 6). If the egg nucleus had not fused (because there were no sperms in its half of the egg), development took place in the sperm closest to the ligature in the other half. This sperm nucleus and the egg nucleus moved towards each other, and met on, or close to, the bridge. Fusion of the pronuclei then took place. Fankhauser concluded from these observations,

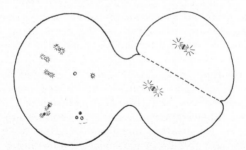

Fig. 6. Egg of a newt, **Triton palmatus,** constricted before the beginning of cleavage (cf. Fig. 13). In the half on the right, the egg nucleus has fused with a sperm nucleus, and cleavage has set in. In the left half, the sperm nuclei farthest removed from the constriction are dividing, whereas those closer to it are inhibited. After Fankhauser.

(1) that there is an attraction between male and female pronuclei, (2) that the egg nucleus promotes the development of the nearest sperm, and (3) that this "favoured" sperm, and later the zygote nucleus originating from the fusion of the pronuclei, inhibit the development of the remaining sperms. It seems probable that these promoting and inhibiting influences are due to substances secreted by the nuclei, and diffusing into the protoplasm.

Some observations on the egg of the fresh water snail *Limnaea* suggest that the mutual attraction between egg and sperm nuclei does not arise until both pronuclei are swollen. This swelling itself depends upon a special condition of the egg cytoplasm, a condition which normally is not realised until after the second maturation division, but which may be precipitated by external influences (Raven and Roborgh, 1949).

The penetration of the sperm into the egg thus starts a series of processes in each of which the components of egg and sperm interact, and which result in the initiation of the development of the fertilised egg. Now it is a very remarkable fact that the same result can be achieved along completely different lines. In eggs of animals belonging to several groups, including the mammals, development has been successfully provoked by means of a wide variety of artificial treatments. This is called *artificial parthenogenesis*. Evidently, the stimulus for development, normally given by the penetrating sperm, is not very specific; very different external stimuli may have the same result.

In the sea urchin egg, for instance, a method devised by J. Loeb gives very good results. Unfertilised eggs are first treated for a few minutes with a weak solution of butyric acid in sea water. When returned to normal sea water, they immediately form fertilisation membranes. After 15-20 minutes they are then brought, for a period of 30-60 minutes, into sea water made hypertonic by the addition of sodium chloride. Finally they are returned to normal sea water. Many of the eggs then cleave, and develop into normal larvae.

In the eggs of the frog satisfactory results are obtained with a treatment discovered by Bataillon (1910-1913). When punctured with a fine needle, such eggs extrude their perivitelline fluid, and develop a fertilisation membrane. Cleavage, however, occurs only if the eggs have previously been moistened with blood. It appears that, for normal development to occur, a live blood corpuscule, containing a nucleus, must be brought into the egg. In this way, cleavage can be provoked in great numbers of eggs, and in a certain percentage of the cases normal further development will take place. A number of parthenogenetic larvae have even been raised successfully through metamorphosis.

The experiments on artificial parthenogenesis have provided the opportunity for a further analysis of the processes that initiate development. Two components of this phenomenon can now be separated experimentally, which in normal fertilisation are so interwoven that they can hardly be distinguished. For instance, sea urchin eggs have been treated with butyric acid, as in the first part of Loeb's method, but then kept in normal sea water. When the eggs are brought back into normal sea water, the cortical reaction takes place, with the result that the fertilisation membrane is elevated. However, cleavage and further development fail to occur. After some time, the egg nucleus swells, its nuclear membrane disappears, and a single aster appears in the surrounding cytoplasm. Each chromosome divides into two halves, but the halves remain together, and finally all the chromosomes reunite into one nucleus in the centre of the now very much enlarged aster. This process may repeat itself several times, but irregularities soon appear, and, finally, this so-called *monastral cycle* terminates in the death of the egg.

In frog eggs, which have been pricked, but not inoculated with cell material, we again find a cortical reaction with elevation of a fertilisation membrane, followed by a monastral cycle, which in the end comes to a stop, culminating in the death of the egg.

Evidently, certain stimuli release a chain of reactions in the egg, called the *activation* of the egg. The most conspicuous links of this chain are the elevation of the fertilisation membrane, and the monastral cycle. These processes, however, do not result in normal development, but lead into a blind alley, ending with the death of the egg. For normal development to occur, other factors must bring the egg cytoplasm into such a condition that, instead of the monastral radiation, a dicentric radiation, or amphiaster, can be formed. The latter leads to normal division, and thereby to cleavage of the egg. Bataillon has named this process *"regulation"*. In Loeb's method, the treatment with hypertonic sea water is the regulating factor; in Bataillon's method the inoculation of cell material acts as such.

Consideration of these two groups of processes shows that widely divergent means can be used for the activation of the egg in different species. Both mechanical stimulation (pricking) and physical treatment (illumination, induction shock, increase or decrease of temperature, or of osmotic pressure) will serve the purpose. Chemical treatment especially, however, has been applied on a large scale (various acids, salts, and alkalis, but also non-electrolytes, such as urea, saponin, etc.). This variety of methods tends to show that none of them constitutes the specific, natural agent itself. The inherent properties of the egg itself must be responsible for the fact that it reacts in the same way to such entirely different stimuli. Now the results of several investigations indicate that calcium ions play a special role in activation. In several species, the egg can be activated only in the presence of these ions in sufficient quantity (Pasteels, 1938; Moser, 1939). Heilbrunn (1937) holds the view that the egg cortex consists of a protein-calcium compound, which can be broken down by the action of various stimuli. The liberated calcium is then taken up by the internal cytoplasm of the egg. This causes an increase in the viscosity of this region, while the cortex simultaneously liquefies. These processes are considered to be the real cause of activation.

Membrane formation and monastral cycle, the two processes especially characteristic of activated eggs, are not linked together inseparably. In the starfish, the formation of a monaster unaccompanied by a fertilisation membrane can be caused by treatment of the egg with a mixture of certain chlorides (Dalcq). On the other hand, in certain cases membrane formation may take place without a monastral cycle. Nor is

the fertilisation membrane indispensable for further development of the egg. It is possible to prevent membrane formation
in eggs, either after fertilisation, or during parthenogenetic
development. This does not always upset development.

Activation causes a marked change in the physiological
properties of the egg. A great increase in the *permeability* of the
egg surface is quite common. This is probably due to the cortical
reaction. In the sea urchin egg, many chemical reactions take
place in the first few minutes after activation. Large molecules
are first broken down into smaller ones, and, somewhat later,
this is followed by resynthesis. Finally, activation often results
in infertilisability of the egg. This, however, is only true in the
case of optimal activation; if the activating agent has operated
for too short or too long a period, so that activation is incomplete, the egg retains its fertilisability.

In normal fertilisation, the egg is activated by the penetrating sperm, as can be seen from the occurrence of the cortical
reaction. Runnström (1952) assumes the following hypothetical
sequence of events to occur in normal fertilisation. Through the
activity of an enzyme present in the spermatozoön a breakdown
of a polysaccharide present in the cell surface occurs. In this
breakdown an acid is formed. Under the influence of a shift in
the pH of the cortex, calcium ions are liberated which cause an
activation of proteolytic enzymes. These in their turn act on
particulate inclusions of the egg releasing various substances.

Apparently the sperm introduces a "regulating" factor
as well, preventing the occurrence of a monastral cycle, and
allowing the nuclei of egg and sperm to join in the formation
of a bipolar mitosis, which leads to cleavage of the egg. We
have seen above that the amphiaster usually arises in connection with the development of the sperm aster, viz. in the central
liquefaction area of the latter.

According to the classical view, an important role in cell
division is played by a specific component of the cell, the
centrosome or *cytocentre*. This has the capacity to divide, and
its halves then form the poles of the new mitotic spindle. Now,
Boveri assumed that the egg cytocentre after the completion
of the maturation divisions has lost the capacity for division.

The sperm, however, contains a cytocentre in its middle piece. This acts first as the starting point for the formation of the sperm aster in the egg cytoplasm. Next it divides, and an aster develops around each half; these asters then form the poles of the cleavage amphiaster. Development is thus initiated by the activity of the cytocentre carried by the sperm.

Later this view was questioned. For some time the reality of the cytocentre as a general organelle with a special function in cell division was even doubted. But recent observations by means of electron microscopy have shown that it exists in cells of various organisms, and contains a corpuscle of characteristic structure, the *centriole,* which apparently is selfduplicating, and plays a leading role in cell division, especially in the formation of the asters and spindle.

It seems hardly doubtful that the centrioles at the poles of the cleavage amphiaster in normally fertilised eggs are derived from the sperm centriole.

However, in artificially activated eggs several asters, so-called *cytasters,* may be produced in the absence of a sperm. Recent electron-microscopic observations have shown that they contain centrioles of characteristic structure in their centre, but it is not clear whether these centrioles have been formed *de novo* in the cytoplasm or by division of the egg centriole. Anyhow, Boveri's view, though explaining the formation of an amphiaster after normal fertilisation, gives no explanation of the occurrence of a bipolar spindle in parthenogenetic eggs.

Therefore other hypotheses have been advanced concerning the nature and localization of the principle of bipolarity.

Dalcq (1928) suggested the following hypothesis. The sperm nucleus, especially its nucleoplasm, is the carrier of the principle that causes bipolarity. The formation, on the disappearance of the nuclear membrane, of a bipolar spindle with its asters, is an inherent property of this nucleoplasm. Dalcq's view is founded on the following observations. (1) Sometimes, in the act of penetrating, the sperm remains stuck with its head in the cortex. In such cases, normal development of the sperm nucleus does not take place. The egg is activated only, and a monastral cycle ensues. This occurs in some cases of heterogeneous fertilisation, e.g. when eggs of the toad *Bufo calamita* are fertilised with sperm of the newt *Triton alpestris* (Bataillon). It may also occur when the sperm has been damaged, e.g. by trypaflavin poisoning (Dalcq). (2) If the fusion of the pronuclei is prevented, each may, under certain circumstances, develop separately in such a way that the male pronucleus undergoes a normal mitosis, whereas the female pronucleus gives rise to a monaster.

This was observed, e.g., by Ziegler (1898) when, in his experiments with fragmented sea urchin eggs, the nuclei of egg and sperm were isolated in different fragments. Wilson (1901), by preventing the fusion of the pronuclei in sea urchin eggs by means of an ether treatment, observed the same phenomenon (Fig. 7). In Amphibia, the same thing may occur in unripe or overripe eggs, in which the fusion of the pronuclei is retarded by the abnormal

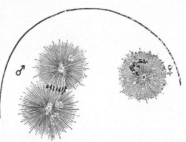

Fig. 7. Egg of a sea urchin, treated with ether. The pronuclei have not fused. The egg nucleus (♀) forms a monaster, the sperm nucleus (♂) an amphiaster. After Wilson.

condition of the protoplasm (Bataillon and Tchou Su, 1934). (3) The co-operation of the *chromatin* of the sperm nucleus is not indispensable for a normal bipolar mitosis. This is proved by some cases of heterogeneous fertilisation, in which the sperm penetrates into the egg, and male and female pronuclei unite, but the chromatin of the sperm nucleus remains entirely compact. In the course of a mitosis which in all other respects is normal, this chromatin does not divide into chromosomes, and is soon afterwards extruded into the cytoplasm, and resorbed. This occurs, e.g., in the case of fertilisation of sea urchin eggs with sperm of the mussel *Mytilus* (Kupelwieser, 1908) (Fig. 8). The same may happen after fertilisation with sperm damaged by radium irradiation (O., G., and P. Hertwig) or trypaflavin treat-

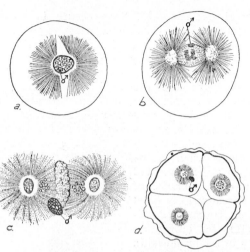

Fig. 8. Egg of a sea urchin, fertilised with sperm of a mussel, *Mytilus*. (*a*) fusion of the pronuclei; (*b-c*) first cleavage, the male chromatin (♂) is eliminated; (*d*) four-cell stage, male chromatin in one of the blastomeres. After Kupelwieser.

ment (Dalcq). (4) Cytasters, produced e.g. in sea urchin eggs, by treatment with hypertonic sea water, are usually unable to divide, and therefore remain monocentric. If, however, they are situated in the neighbourhood of a normal mitosis, they can sometimes "capture" part of the nuclear material of the latter. Thereupon they divide, and form an amphiaster (Fry, 1925). Working with sperm treated with trypaflavin, Dalcq obtained amphiasters which, as a consequence of irregularities in the course of division, contained no chromosomes, but only part of the *nucleoplasm* of the sperm nucleus. These were able to divide repeatedly, thereby giving rise to new amphiasters (Fig. 9).

Dalcq concluded from these experiments that the power of division, especially that of forming a bipolar mitosis, is an inherent property of the sperm nucleus, in particular of its nucleoplasm. Yet there is an objection to this conclusion. There are several exceptions to the rule that co-operation of the male pronucleus

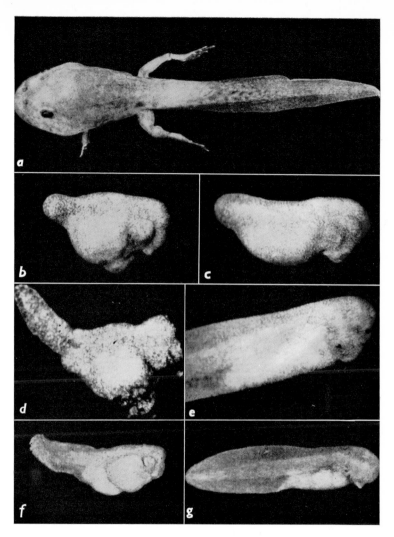

PLATE III. Developmental capacity of nuclei gastrula cells. (a) Meta-morphosing frog tadpole derived from an enucleated egg injected with a presumptive chorda-mesoderm nucleus of a late gastrula. (b-c) Embryos of 4 days, derived from enucleated eggs injected with endoderm nuclei of a late gastrula donor; b. abnormal, c. normal development. (d) Same embryo as shown in figure b, photographed at age of 6 days, illustrating maximum development displayed by embryos of this type. Compare with normal embryo of same age (e). (f) Experimental embryo, age 8 days, showing deficiencies of the same general type as displayed by the embryo in d, but less severely expressed. Compare with normal embryo of same age (g). After Briggs and King (1957) (a, after T. J. King and R. Briggs, **Jour. Embr. exp. Morph. 2, 1954**).

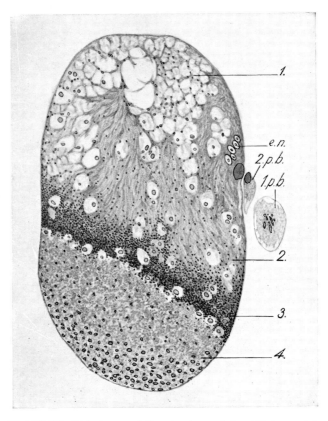

PLATE IV. Disturbance of the structure of the egg by centrifuging. Egg of a snail, **Limnaea stagnalis,** centrifuged before the
beginning of cleavage. *1*: Cap of fat with large vacuoles at the
centripetal pole. *2*: clear cytoplasm with few inclusions. *3*: zone
of mitochondria. *4*: centrifugal cap containing the protein yolk.
The animal pole is at the right, with the egg nucleus (**e.n.**) and
the two polar bodies (**l.p.b.** and **2.p.b.**).

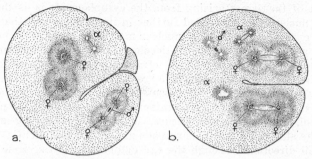

Fig. 9. Two sections of a frog's egg, fertilised with two trypaflavin treated sperms. ♀ : haploid nuclei, originating from the egg nucleus; ♂ : sperm chromatin has remained compact; α: mitotic spindles without chromosomes. After Dalcq.

is necessary for the formation of a normal bipolar mitosis. In *Crepidula* (Conklin, 1904) and *Ciona* (Duesberg, 1926), the female pronucleus will form an amphiaster, even if fusion of the pronuclei is prevented. On the other hand, in Fankhauser's experiments already described (see above, p. 19), the superfluous sperms in the half without the egg nucleus often form a monaster, although, naturally, the nucleoplasm of the sperm is present here. Further, Dalcq's hypothesis does not explain "regulation", such as is caused, e.g., by treatment of the sea urchin egg with hypertonic sea water in Loeb's method. It is true that in Bataillon's method for artificial parthenogenesis in Amphibia regulation is caused by the introduction into the egg of a blood corpuscle with its nucleus, but this nucleus does not unite with the egg nucleus. Moreover, Einsele (1931) has shown that the same result may be obtained with an extract of blood or sperm. According to a later investigation by Shaver (1953), the activity of such extracts is mainly due to a fraction consisting of cytoplasmic granules, which are very rich in ribonucleic acids (mitochondria and microsomes).

For these reasons, Bataillon has attempted an entirely different explanation of the origin of bipolarity. In his opinion, the occurrence of a monastral cycle in activated eggs is due to an "anachronism" between the development of the nucleus and that of the asters arising from the cytoplasm. It is as though the nucleus has proceeded further in its development than has the cytoplasm so that the nuclear membrane disappears before the aster has had time to develop into an amphiaster. In normal fertilisation, the penetrating sperm accelerates the activity of the cytoplasm. This "regulates" the phase difference between nucleus and cytoplasm. Bataillon points out that, in the normal course of fertilisation, first a gelation of the protoplasm, the sperm aster, occurs, with the penetrating sperm as its starting point. This spreads as an "onde de gélification" in all directions through the egg cytoplasm. At the same time the centre of the sperm aster solates (liquefies) again, and it is only then that the amphiaster forms in this central area, and that the nuclear membranes of the pronuclei, which have in the meantime fused, disappear. In Bataillon's opinion, the passing of this "onde de gélification" prepares the egg cytoplasm for the formation of an amphiaster, and thereby abolishes the anachronism between nucleus and protoplasm. In artificial parthenogenesis in Amphibia, the "onde de gélification" starts from the cell fragments with which the egg has been inoculated when it was punctured. Chambers (1921) has shown that in Loeb's method the treatment of the sea urchin eggs with hypertonic sea water causes an "onde de gélification" in the cytoplasm around the nucleus. Therefore, Bataillon's hypothesis is able to explain both the phenomena of normal fertilisation, and those of artificial parthenogenesis.

Special relationships are found in many cases, where the maturation divisions do not take place before the egg is activated, as in various annelids and molluscs. Here as a rule cleavage does not occur in fully-activated eggs, but only when activation is incomplete and polar body formation is suppressed. In these cases the presumptive maturation spindle may either be converted directly into a cleavage spindle, or so-called "submerged" maturation divisions (without extrusion of polar bodies) may occur followed by fusion of nuclei and association of cytocentres. Apparently, the egg cytocentre under these circumstances does not lose its capacity for division.

The various observations and theories may perhaps be summarised as follows: the formation of a bipolar cleavage spindle in normal fertilisation is due to the activity of the sperm cytocentre. The division of a cytocentre is in general somehow dependent on the presence of a nucleus, especially of its nucleoplasm. After extrusion of the polar bodies the egg cytocentre is, as a rule, unable to divide, but it may regain this capacity under certain circumstances, e.g. when an "onde de gélification" has passed through the cytoplasm.

CHAPTER III

The structure of the fertilised egg

The structure of the egg at the starting point of its development seems to be quite simple. But, as we have seen in the Introduction, it is conceivable that this simplicity is only apparent. The whole complicated spatial structure of the later organism might already be present, in a not easily recognisable form, in the egg. According to the *preformation* theory, which had many supporters in the 17th and 18th centuries, the complete young animal was already contained in the fertilised egg, in the same way as a complete stalk with leaves, flowers, etc., can be contained in the bud of a plant. The development of the egg would be no more than the unfolding of this preformed young germ. The opposite view was maintained by the theory of *epigenesis*, which held that the embryo was not yet present, as such, in the egg, but that it would arise as a new product in the course of development. In the 19th century, a thorough study of the developmental phenomena became possible, thanks to the great progress in microscopic technique made in that period. By this means, the dispute between preformation and epigenesis was at that time settled in favour of the latter theory. A few years after 1880, however, the issue was reopened, in a different form, by Wilhelm Roux. It had indeed been shown that the embryo as such is certainly not yet present in the fertilised egg, but was it not possible that the "degree of multiplicity" of the egg's spatial structure was almost, or quite as high as that of the embryo? The egg might contain, in an invisible form, such a complex system of causal factors, that each of its parts would be able, quite independently, to produce one definite part of the embryo, and nothing else. This implied that each part of the embryo would be preformed in one definite part of the egg. As a consequence of the structure of the egg, in which each part had its fixed place, the various organs, though each developing independently, would later fit together as parts of a "mosaic". Therefore, development would not involve an "increase of spatial multiplicity", but only the manifestation of a previously invisible, preformed, spatial multiplicity. This is the hypothesis of *neo-preformation*. Its alternative, that of *neo-epigenesis*, is the hypothesis that the spatial multiplicity of the embryo is in no way preformed in the egg, but that it arises during development.

W. Roux, the creator of the science of "developmental mechanics", considered the solution of this problem its first and most important task. Since then, it has been the motive for a great number of investigations in a wide variety of animal groups. Such studies are characteristic of the first period of research in developmental physiology. On the strength of these results, we can now assert on good grounds that the spatial structure of the egg is generally very simple, and that therefore the spatial multiplicity of the embryo increases strongly during development, as is postulated by the neo-epigenetic hypothesis. We shall now discuss a number of experiments which have led to this conclusion.

We have already seen that the first step in development is the cleavage of the fertilised egg into two cells. Each of these cells divides into two again, and so on, until a great number of cleavage cells, or blastomeres, has been formed. This process does not involve cell growth, so that the resulting blastula is still of about the same size as the original egg. In 1891, Driesch made an experiment which was to prove of the greatest importance for the development of our views in this field. He took sea urchin eggs which had just begun developing, and were in the two-cell stage. Driesch shook those eggs in a test tube with sea water, and thereby separated the two blastomeres of one egg in a number of cases. He then studied the further development of these isolated halves, and observed that each of them was able to develop into a harmoniously built young larva. These larvae were only half the size of ordinary ones, but were otherwise normal in form (Fig. 10). Moreover, Driesch found

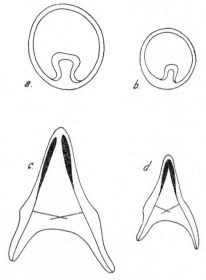

Fig. 10. Gastrula and pluteus of a sea urchin. Left: from a whole egg; right: from a ½-blastomere. The latter are smaller, but harmoniously built. After Morgan.

that even blastomeres isolated at the 4-cell stage could still develop into harmoniously built larvae of reduced size.

The same phenomenon, the origin of normal embryos from one half or one quarter of an egg, was found to occur in other groups of animals as well. In several species of newts, for example, the two blastomeres of the two-cell stage can be separated by constricting the egg with a hair ligature in the cleavage plane. Here, too, under certain circumstances both halves will develop into harmoniously built embryos of half the normal size, which lie together in one egg capsule (Spemann, 1903), (Plate I). Even if newt eggs are divided into two halves at a much older stage, after the completion of cleavage, each half may still develop into a complete, normal embryo.

From other eggs, large parts of the egg cytoplasm can be

removed before the beginning of
cleavage, without any disturbance
of their further development.

In such cases, a normal embryo
originates from part of an egg.
However, the opposite, development of a single embryo from
the fusion of two or more eggs,
has also been observed. Driesch
(1900) has done this experiment
in sea urchins, and it has been repeated and extended by Bierens
de Haan (1913). The envelopes of
two sea urchin eggs were removed,
and the eggs were gently pressed
together. They then often fused
into one germ of double size. Here
too, if certain conditions are fulfilled, a single, harmoniously built
embryo will develop from the
product of such a fusion (Fig.
11).

Mangold and Seidel (1927) have
done the same experiment in newts.
After removal of the envelopes,

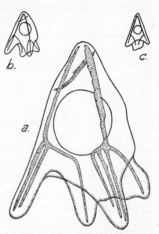

Fig. 11. (a) pluteus of a
sea urchin, **Paracentrotus
lividus**, which has originated
from the fusion of two eggs.
(b) the same pluteus, for
comparison with (c) a
normal pluteus drawn at the
same scale. After Bierens
de Haan.

newt eggs at the two-cell stage often assume a dumb-bell shape.
When one was placed crosswise on top of the other, two of
these dumb-bells often fused into one single germ. Each pair
of opposite quadrants of such an embryo was derived from
one of the original eggs. Part of the products of these fusions
proved to be able to develop into single, more or less harmoniously built embryos. This was possible even when the
fused eggs belonged to different species of newts so that the
resulting embryo was a "chimera", consisting of quadrants
which alternately belonged to either of the two species
(Plate II).

The view that the egg has a fixed spatial structure, and that
this is responsible for the development of its parts into definite

organs, is rendered highly improbable by these experiments. It becomes even less tenable in the light of the following facts. A profound disturbance of the structure of the egg cytoplasm often does not result in aberrant development. Such a disturbance can be caused by centrifuging the egg. Under the influence of the centrifugal force, the materials of the egg are arranged according to their specific gravity. The lighter substances (especially the fats) are accumulated at the centripetal pole, the heavier material moves toward the centrifugal pole. In this way the contents of the egg are arranged in layers, an arrangement very different from the normal structure of the egg cytoplasm (Plate IV). Nevertheless normal development will often take place in such centrifuged eggs, e.g. in polychaete worms and in molluscs. In the course of this development, the stratification is lost to some extent, and the material of the egg is redistributed more evenly. Finally, normal embryos will be formed.

All these experiments give positive indications that the cytoplasm of the fertilised egg cannot contain a complex spatial structure, in which each part represents one definite organ, or part of the body, of the future embryo. For if this were true, the division of such a structure into two halves (as in the experiments with isolated blastomeres) would be bound to cause defects in the embryos developing from each half. Further, it is very difficult to imagine how fusion of two of these systems could again result in a unity of entirely similar properties. Finally, one cannot understand how such a structure, once it has been disturbed (as in the centrifuge experiments), could return to its original state.

The conclusion obviously must be that the egg cytoplasm has no such complicated structure, no "extensive multiplicity", but that it is a more or less homogeneous system. Such a system — a water drop is a convenient model — can be divided, and, alternately, two such systems can fuse, without any change of properties. A change which produces heterogeneity will be undone as soon as the factors which caused it suspend their activity, because the components of the system return to their position of equilibrium, i.e. the homogeneous distribution.

Therefore, the multiplicity of the egg cytoplasm must be mainly intensive.

For the sake of completeness, an alternative explanation of the experiments described above must be mentioned. One can still maintain that the egg cytoplasm has a complicated structure, which is disturbed in these experiments. In that case, the existence must be postulated of a more or less mysterious "vital force" which sees to it that the normal structure of the egg is restored after the disturbance. Such an explanation was given by Driesch, who regarded the result of his experiments as proof of the existence of an *entelechy* which, after a disturbance, "regulates" the course of development again. Undoubtedly, this is a possible explanation, but it is not the most obvious one because it is founded on the introduction of a new, and entirely hypothetical, factor. Now it is the task of science always to look for the simplest possible explanation, and to accept this as long as it has not been disproved. Therefore, we must reject Driesch's hypothesis so long as no more compelling arguments can be advanced in favour of it, and we must accept the conclusion that, on the whole, the egg cytoplasm is a homogeneous system with intensive multiplicity only.

In the Introduction it was pointed out that the various animal groups by no means behave identically in their development. This applies also to the results of the experiments described above. Often they show different results in other eggs than those mentioned. Centrifuge experiments, for example, do in fact cause disturbances in the development of the eggs of many species. For some time past, therefore, two types of eggs were distinguished, (1) "mosaic eggs" which possessed a complicated spatial structure, and in which each disturbance of the system resulted in disturbed development, and (2) "regulation eggs", with a poorly developed, or highly plastic spatial structure, in which disturbances of the system were easily "regulated". This distinction, however, has proved unfounded. It has been shown that the aberrant behaviour of "mosaic eggs" is not due to fundamental differences in the structure of the egg, but to a number of adventitious phenomena (cf. p. 69). Therefore, it can be said of these eggs as well that, *at the beginning of development, the egg cytoplasm has no, or at most very little, spatial multiplicity.*

The application of modern methods of investigation has considerably increased our insight into the structure of the cytoplasm in the last few years. In some respects, it was shown to have the properties of a liquid. The viscosity of the protoplasm of sea urchin eggs is only a few times higher than that of water. In many eggs obvious protoplasmic currents can be observed. Brownian movements of granular inclusions have been seen in eggs, e.g. of nematode worms. Other observations, however, show that cytoplasm is not a pure liquid, but that it has, for instance, elastic properties. Iron particles taken up by a cell can be moved by means of a magnetic field, but will elastically return to their original position when the field is removed.

According to Frey Wyssling protoplasm contains a network of long polypeptide chain molecules, interconnected by side-chains. These "contacts", however, are continually being broken and re-connected. The meshes of the network are filled with a watery solution of salts, and with lipids, phosphatides, etc. These substances are located, in variable arrangements, around the free side-chains of the polypeptide molecules.

In recent years, electron microscopy has enabled us to study the submicroscopic structure of the cytoplasm. It has been shown to contain various components of characteristic structure, only some of which are visible with the light microscope. Among the latter, we may mention the *mitochondria,* which were known for a long time as small granules in the cell, stainable by special methods. In electron microscopy of ultrathin sections, they present a characteristic structure. They are lined externally by a double membrane, from which coulisse-like partitions, the so-called mitochondrial crests, extend into the interior. We know from biochemical investigations that many cell enzymes, especially those enzymes catalysing the oxidative breakdown of combustible substances and the phosphorylations leading to the formation of high-energy phosphate compounds, are bound to the mitochondria, which therefore play a predominant role in the generation and transmission of energy within the cell.

Other enzymes, especially acid phosphatase and similar hydrolytic enzymes, are contained in little vesicles, the *lysosomes*. They may play a part in various breakdown processes in the cell.

Another regularly occurring component of the cytoplasm is a complex consisting of stacks of lamellae, forming the walls of flattened pouches, and surrounded by numerous small vesicles and larger vacuoles. They are thought to correspond partly to the structures called *"Golgi apparatus"* by light microscopists, and to play a role in the accumulation and perhaps transformation of substances synthesised by the cell.

Further, most cells contain a structure, called *"endoplasmic reticulum"* by Porter, but actually consisting of a system of cisternae and pouches lined by double membranes. These membranes may be covered on their outer side by innumerable small granules, about 150 Å in diameter, the *ribosomes*. In that case the endoplasmic reticulum probably corresponds to the *ergastoplasm* of older authors. Besides, free ribosomes, not bound to membranes, may be found in the cytoplasm. The ribosomes, which are very rich in ribonucleic acid, are thought to be the main centres of protein synthesis in the cell.

Apart from these organelles, which are common components of most cells, the egg cells as a rule contain as further inclusions the globules and droplets belonging to the yolk. They are bodies rich in carbohydrates, lipids and proteins, of which various kinds may be present in the egg cell.

If we study the arrangement of these cell organelles and inclusions in the uncleaved egg, it appears that they are in general uniformly distributed. Distinct localisations of these components are at this stage scarcely encountered.

At the egg surface, we find a layer of a more solid consistency, the *egg cortex*. In some cases, as in sea urchin eggs, it is a few microns thick, and consists partly of a gelated layer, which is responsible for most of its mechanical properties. In other cases it is thinner and has another structure. In electron microscopy, its only consistently observed component appears to be an extremely thin double membrane at the surface, about 100 Å thick. This *plasmalemma*, which probably consists mainly of proteins and phospholipids, may be responsible for some of the physiological properties of the cell surface, e.g. its permeability. Moreover, there are indications that the morphogenetic role of the egg cortex (cf. below, p. 52) is, at least in some eggs, also due to the structure of this membrane.

All this tends to show that the composition of the egg cytoplasm is very complicated indeed, but that its structure does not adumbrate the structure of the future embryo. It is mainly in the nature of an intensive multiplicity, as defined above (p. 3).

Apart from the egg cytoplasm, however, the fertilised egg also contains a nucleus, originating from the fusion of the nuclei of egg and sperm. During cleavage this nucleus divides into a number of cleavage nuclei which find their way into the various blastomeres. Could it not be that the egg has a complicated spatial structure, localised, however, not in its cytoplasm but in the nucleus, in which, after all, the spatial multiplicity of the future embryo might be preformed in some way or another?

Weismann advocated this hypothesis in 1892. In his opinion, the nucleus of the fertilised egg contained the *id*, a three-dimensional structure, consisting of material particles, the *determinants*. In the course of the nuclear divisions during cleavage the id would be distributed over the various cleavage nuclei. The nucleus of each blastomere would receive only part of the determinants, and, eventually, there would be only one determinant in each cell. This would then exert its influence upon the cell, thereby unequivocally determining its differentiation. In other words, the spatial multiplicity of the embryo would be preformed in the architecture of the id; the rules governing the distribution of the determinants during cleavage would keep the process of development on its normal course.

Two deductions from Weismann's theory are open to experimental verification: (1) there must be qualitative differences among the nuclei formed during cleavage, and (2) disturbances of the normal course of cleavage must result in an abnormal distribution of determinants over the cells, and therefore in abnormal development. If it can be shown ex-

perimentally that these conclusions do not correspond to reality,
it follows that the theory is not correct.

Now it can already be seen from the experiments on the
isolation of cleavage cells (see above, p. 31) that the nuclei of
the first few blastomeres are not qualitatively different. We have
seen that a single blastomere from the two-cell stage of a newt
or the 4-cell stage of a sea urchin can still produce a complete,
normally built embryo. This proves that the nucleus of these
cells must still be equivalent to the original nucleus of the
fertilised egg. In other words, no division has occurred, up
to this stage of cleavage, that led to non-equivalent results.
This is demonstrated even more clearly by the following
experiments. Driesch (1893) kept developing sea urchin eggs between two sheets of glass, so that they were slightly flattened.
This changed the direction of the plane of the divisions, and

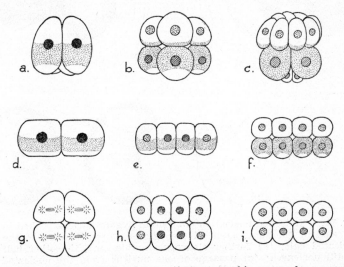

Fig. 12. Diagram of the cleavage in a sea urchin egg under pressure.
(a-c) normal cleavage (4, 8, and 16-cell stages); (d-i) cleavage in
an egg flattened by external pressure; (d-f) seen from the side;
(g-i) seen from above. Analogous nuclei marked in the same way
in all cases. After Dürken.

resulted in abnormal positions of the cleavage cells relative to one another. Consequently, the cleavage nuclei became located in the "wrong" cells (Fig. 12). In spite of this, the eggs developed into normal embryos. It follows that disturbance of the course of cleavage does not influence further development.

Spemann (1928) constricted fertilised, but uncleaved newt eggs with a hair ligature, giving them a dumb-bell shape. The nucleus lay in one end of the dumb-bell shaped egg, and only this part was able to start cleaving. The other, non-nucleated

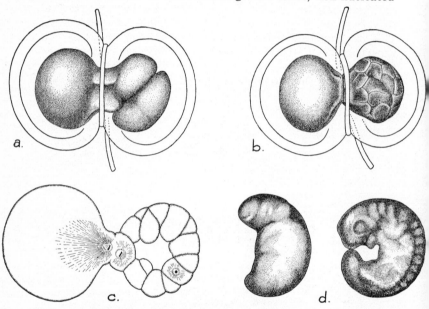

Fig. 13. "Retarded nucleation" in **Triton**. Zygote nucleus pushed to one side by the constriction of the egg. (*a*) first cleavage in the nucleated half; (*b*) further cleavages in this part, the other half is still uncleaved; (*c*) passage of one nucleus from the nucleated into the non-nucleated half, beginning of development in the latter; (*d*) the two embryos that developed from the constricted egg; the one produced by the half in which nucleation was retarded is considerably younger, but normally built. After Spemann and Fankhauser.

half remained uncleaved (Fig. 13 a, b). After a certain number of divisions, however, when the nucleated part had divided into 8 or 16 cells, one of the cleavage nuclei would pass from this half, through the ligatured peduncle, into the other part (so-called "retarded nucleation" of this half), (Fig. 13 c). This half then began to cleave as well. If at this stage the two halves were entirely separated, each would, under certain circumstances, develop into a normal embryo (Fig. 13 d). Therefore, the nucleus which passed through the peduncle, and which represents only one eighth or one sixteenth of the original fertilisation nucleus, is sufficient to bring about normal development of the half in which nucleation was retarded. It is evidently equivalent to the fertilisation nucleus. This proves once again that, at least in the first stages of development, there is no qualitative difference between the division products of the nucleus.

This conclusion could be extended to later stages by the interesting experiments of Briggs and King (1953-57) on nuclear transplantation in amphibian eggs. Unfertilised frog eggs were activated by pricking, then the nucleus was removed. A nucleus from a cell of a later stage, surrounded by some of its cytoplasm, was then injected with a micropipette into the enucleated egg. When the injected nucleus was taken from a cell in the animal half of a blastula or an early gastrula, part of the eggs cleaved normally and developed into normal embryos. This proves that, in the animal half of the blastula and the early gastrula, the nuclei have unrestricted potentialities and are still equivalent to the fertilisation nucleus. In experiments with nuclei taken from a late gastrula, in which differentiation of the cells had begun, the percentage of normally cleaving eggs was much lower, however, and many of them were arrested at blastula, gastrula or abnormal neurula stages. In the abnormal embryos, cellular differentiation was good in some tissues, deficient in others, depending on the region of the donor from which the transplanted nucleus had been taken. Briggs and King conclude from their results that the nuclei lose their equipotentiality and become differentiated during gastrulation. This progressive specialisation of the nuclei goes hand in hand with cellular differentiation; therefore, it cannot be considered as the cause of the determination of the cells, but must rather be a consequence or attendant phenomenon of cell differentiation (Plate III).

These experiments prove that Weismann's theory cannot be maintained. Development is not due to a qualitatively unequal distribution of a spatial system of developmental factors, localised in the nucleus of the fertilised egg.

We must conclude that the spatial multiplicity of the later organism is not preformed as such, either in the cytoplasm or in the nucleus of the fertilised egg. Therefore, development appears to involve an increase in extensive multiplicity.

CHAPTER IV

Polarity and symmetry;
The cortical field

In the previous chapter we have seen, that, in broad outlines, the newly fertilised egg is to be regarded as a homogeneous system, and that its multiplicity is only intensive, not extensive. However, we shall now discuss a number of phenomena that show that this is true only with certain restrictions.

Even a superficial examination shows that the egg can not be so entirely homogeneous as is, for instance, a water drop, because we can distinguish between egg cytoplasm and nucleus. But, apart from this, a further organisation is demonstrable in the egg; this is expressed in its *polarity* and its *symmetry*.

All animal eggs have a polar structure, i.e. two opposite poles can be distinguished, called the *animal* and *vegetative poles* respectively. They are connected by the *main axis* of the egg. Considering the egg as a globe, we may call all planes that contain the main axis *meridian planes*, whereas the plane that bisects the main axis at right angles is the *equatorial plane*. In some cases, e.g. in the oblong eggs of insects (Fig. 26) and cuttlefish, the polarity is revealed in the shape itself of the egg. In other cases, the egg is more or less spherical, but its polarity is evident in other ways: the polar bodies are given off at the animal pole of the egg (p. 10); after fertilisation, the zygote nucleus often lies in an eccentric position, nearer to the animal pole. The first two cleavages nearly always take place in meridional planes so that they intersect in the main axis of the egg.

Further the polarity is often apparent in the arrangement of the inclusions of the egg cytoplasm. In many eggs, the food substances which together constitute the yolk are not evenly distributed; often their density increases from the animal toward the vegetative pole. There may also be differences in pigmentation. In many amphibians, for example, the animal region of the egg surface is darkly pigmented, whereas the vegetative region is unpigmented (Fig. 14).

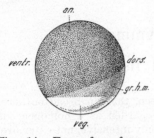

Fig. 14. Egg of a frog, Rana fusca, after the formation of the grey crescent (*gr.h.m.*). *an*: animal side; *veg*: vegetative side; *dors*: dorsal side; *ventr*: ventral side. After Schleip.

In the great majority of animals, the unripe eggs (*oöcytes*) possess a polar structure already on leaving the ovary. Evidently, the polarity originates during the growth of the egg in the ovary. In many cases a connection can be observed between the polarity of the oöcyte, and the way in which it is attached to the ovary. The place of attachment, which is also the place where the food stream from the maternal tissues enters the oöcyte, is in some groups the future animal pole, in others the future vegetative pole of the egg.

The polarity of the egg is related to the main direction of the embryo that develops from it. As a rule, the animal side becomes the front end, the vegetative side the hind end of the embryo.

Apart from polarity, many eggs possess an obvious *bilateral symmetry*. This, too, may be expressed in the shape itself of the unfertilised egg, as in insects and cuttlefish. In other cases, the symmetry is not visible until after fertilisation. In the eggs of many amphibians, a zone of lighter pigmentation, the so-called *grey crescent*, develops on one side of the egg along the boundary of the dark animal half in the first few hours after fertilisation. This marks the dorsal side of the egg, and later of the embryo. The opposite side becomes the ventral side (Fig. 14). Once the grey crescent has been formed, we can divide the egg into symmetrical halves in only one way, viz. through the meridional plane which bisects the grey crescent. In normal development this plane, the median plane, will become the plane of symmetry of the embryo. It contains the main axis, and also the *dorso-ventral axis* which is at right angles to the main axis, and connects two opposite points on the equator of the egg.

It must be remarked, by the way, that a third axis, which is perpendicular to the median plane, represents the left-right direction of the future animal. As most higher animals are not strictly bilaterally symmetric, but more or less asymmetric, one may ask whether this asymmetry is also, somehow or other, expressed in the structure of the egg. There are some indications that such is indeed the case, especially in animals with a pronounced asymmetry, as snails.

By making small lesions in the surface of frog eggs, Ancel and Vintemberger (1935) were able to show that the formation of the grey crescent is due to shifting of the egg cortex relative to the deeper layers of the cytoplasm. On the dorsal side, the egg cortex moves toward the animal pole, and also slightly towards the median plane. Part of the superficial pigment gets caught in this movement, and a less pigmented area, the grey crescent, is left behind.

The ascidian egg is another example of an egg in which bilateral symmetry does not become visible until after fertilisation (p. 63).

For frog eggs, Ancel and Vintemberger (1948) have solved the problem of what determines the position of the plane of symmetry of the egg. Under natural circumstances, the fertilising sperm is the main factor. Its point of entry into the egg determines the plane of symmetry, the grey crescent appearing at the opposite side of the egg. During the first hour after insemination, the plane of symmetry may be shifted at will by experimental means, however. Immediately after spawning, the orientation in space of the amphibian egg is arbitrary. Some time after fertilisation, however, the eggs become freely movable within their membranes. They are now oriented by gravity, and rotate so that the main axis is vertical, with the animal pole pointing upwards. This rotation proved important for the determination of bilateral symmetry. Normally, its action does not suffice to modify the action of the sperm. But

under experimental conditions, if the amplitude of the rotation
or its effectiveness are increased, it may modify or annihilate
the influence of the point of sperm entry. The meridional plane
in which the rotation takes place will then become the plane
of symmetry of the egg. The grey crescent will form on the
side along which the vegetative pole has descended. By fixing
unfertilised eggs so that their animal poles point upwards,
and activating them by means of an electrical induction shock
(p. 22), both the action of the sperm and of the egg rotation can
be eliminated. In this case, the grey crescent may be formed
anywhere. This shows that the occurrence of bilateral symmetry, *per se,* does not depend on the one-sided penetration of
a sperm or on the orienting rotation. But the direction of the
plane of symmetry may be determined by these two factors.
According to Ancel and Vintemberger, they cause a slight
asymmetry in the position of the cortex relative to the
internal parts of the egg. In the case of the rotation, it is the
relative displacement of the membranes with respect to the
egg surface which produces this, so that a rotation of the
membranes relative to an egg remaining in its position of
equilibrium is just as effective as a rotation of the egg within
the fixed membranes. The asymmetry of the cortex then directs
the further processes leading to the formation of the grey
crescent.

Recent investigations by Vintemberger and Clavert (1953-
60) have shown that similar relationships occur in birds. Here
it is the rotatory movement to which the eggs are subjected
during their passage through the oviduct of the female which
determines the position of the embryo in the egg. Since the
eggs have already begun to cleave at this time, the determination of bilateral symmetry takes place here at an even later
stage than in the Amphibia.

We have already said that the phenomena of polarity and
symmetry complicate the very simple picture of the egg drawn
in the previous chapter. They show that it would not be correct
to regard the egg as a completely homogeneous system. Polarity
and symmetry can only be explained on the assumption that
there are local differences in composition within the egg, and

a closer examination of the egg proves the truth of this assumption. Some of its physical and chemical properties appear to be unevenly distributed. Often these distributions have the character of gradient systems.

By *"gradient-field"* is meant a spatial distribution of a physical or chemical quantity, whereby the value of this quantity gradually changes from point to point. The properties concerned are always so-called "scalar" quantities, i.e. a numerical value can be attributed to them (which might be read on a scale), but no direction, e.g. temperature, pressure, concentration of a substance, electrical potential, and the like. In a gradient-field, we can distinguish planes of equal intensity. At all points in such a plane the scalar quantity has the same value. The rate of change of the quantity at any given point in the field is called the gradient at that point. At each point of the field, the gradient has one, and only one, definite value and direction, since it is always measured at right angles to the plane of equal intensity through the point.

Now in many eggs gradients have been found which show some connection with polarity or symmetry. We have already mentioned that there is an increase in yolk concentration along the main axis of many eggs (p. 43). This may be called a yolk gradient. Further, Child and his collaborators have for many years stressed the importance of local differences in the intensity of metabolism, as expressed in differential sensitivity to noxious influences in different regions. In animal eggs, as a rule, the sensitivity is greatest at the animal pole, and from there it decreases gradually in the direction of the vegetative pole. This points to a variation in metabolism of the egg along its main axis. Similar results have been obtained in determinations of the redox potential (rH) of the egg cytoplasm with various dyes, such as Janus green and methylene blue (Ries and Gersch, 1936).

In this context experiments by Spek (1930-1934) deserve mention. He measured the degree of acidity (pH) of protoplasm by means of indicator dyes (substances whose colour depends on the pH of their environment), e.g. neutral red, nile blue hydrochloride and brilliant cresyl violet. Initially, most eggs showed a more or less uniform coloration, but shortly after the beginning of development a "bipolar differentiation" took place, i.e. the colours of animal and vegetative regions began to differ. The colour of the animal cytoplasm indicated an alkaline reaction, that of the vegetative part of the egg an acid reaction. Spek explained this as follows. Originally colloid particles with positive and negative electrical charges were mixed throughout the egg. Later, the positive particles move to one pole, and the negative particles migrate to the other pole. The mixture is thereby separated, and particles of the same sign become concentrated at each of the poles. This migration of the particles was assumed to be caused by an electric field, due to differential penetration of certain ions (potassium ions playing a major role) from the environment into the egg. A difference in permeability of the egg surface near the animal pole was supposed to be responsible for this penetration. Later investigations, however, have shown that "bipolar differentiation" is due not to segregation of colloid particles of opposite signs in the cytoplasm, but rather to shifting of the yolk of the egg. This material, consisting of fat and protein globules, now assumes its final position with regard to the egg axis (Raven, 1938b). The acid yolk proteins accumulate at the vegetative pole; their disappearance from the animal side causes the pH there to shift in the direction of alkalinity.

Gradient systems play an important role, for instance, in the development of sea urchin eggs. Cleavage in these eggs is very regular. It results in a blastula, a vesicle with a wall consisting of a single layer of cells with cilia on their outer surfaces. In the next stage, the wall of the blastula invaginates at the vegetative pole, thereby forming the *archenteron* of an embryo called a *gastrula* (Fig. 31). At the animal pole the gastrula possesses a long tuft of cilia (Fig. 15a). The further development of a gastrula into a *pluteus larva* involves the growth of long "arms", and the secretion of a skeleton of calcareous rods by cells originating from the vegetative wall of the blastula. Moreover, the larva is now surrounded by a band of cilia which also extends over part of the arms (Fig. 15b).

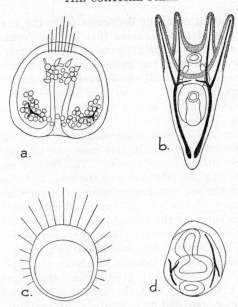

Fig. 15. Sea urchin larvae. (*a*) normal gastrula and (*b*) normal pluteus; (*c*) embryo with a large tuft of cilia, produced by a fragment of the animal side of an egg; (*d*) embryo containing gut and skeleton formed by a fragment of the vegetative side. After Hörstadius.

Driesch has shown, as mentioned above (p. 31), that sea urchin blastomeres isolated at the 4-cell stage can still develop into a complete larva. This stimulated several investigators, in particular Hörstadius (since 1931) to study the developmental potencies of blastomeres and groups of blastomeres, isolated at later stages. In these investigations, the potencies of the egg material were shown to change gradually from the animal towards the vegetative pole. Isolated animal halves of a germ will produce a larva with a large apical tuft, but without gut and skeleton; isolated vegetative halves will grow into a larva with a large gut, and an irregular skeleton, but without either apical tuft or band of cilia (Fig. 15). Evidently, the potency for the formation of a gut decreases from the vegetative towards the animal pole, whereas that for the formation of an apical tuft decreases in the opposite direction. When cell groups from different parts of the germ are brought together, more or less normal larvae will result even from highly abnormal combinations of blastomeres, so long as cells with animal and cells with vegetative potencies are present in balanced proportions.

Runnström (1928) has suggested that there are two oppositely directed gradients in this case, one animal-vegetative, and one vegetative-animal gradient. The fate of each cell will be determined by the interaction of the two. Addition of certain chemicals to the sea water surrounding the germ may produce an ascendancy of one of the two gradients over the other. Lithium ions suppress the animal gradient so that the vegetative differentiation of the embryo begins to preponderate at the expense of the animal differentiation (Fig. 16). Other substances, such as sodium thiocyanate, have the opposite effect; they reinforce the animal gradient at the expense of the vegetative gradient so that the animal differentiation of the embryo is promoted, and its vegetative differentiation inhibited.

If the polarity of the egg is determined by axial gradients, disturbance of these gradient systems may be expected to cause changes in polarity. We have mentioned above (p. 34) that the structure of the egg can be profoundly modified by centrifuging. As a rule, the proteid yolk is heavier, and the fatty yolk lighter than the clear cytoplasm. Therefore, they will accumulate at opposite poles, with the cytoplasm occupying the middle zone of the egg (Plate IV). The direction of this stratification need not coincide with the original polarity, but may be at any angle to it. In this way, the normal yolk gradient and the correlated gradients of metabolism are completely destroyed. Nevertheless, the further development of such eggs shows that in the majority of cases their polarity has remained unchanged. The egg material, displaced by centrifuging, often after a short time returns to its normal position with regard to the original egg axis (Raven, 1938; Raven and Bretschneider, 1942); polar bodies are given off at the original animal pole;

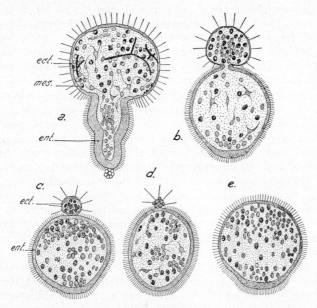

Fig. 16. Abnormal sea urchin embryos, produced by treatment with lithium. (*a*) exogastrula; (*b-e*) extension of the endoderm at the expense of the ectoderm; *ect.*: ectoderm; *ent.*: endoderm; *mes.*: mesenchyme cells. After Herbst.

cleavage and development of the embryo are oriented with regard to the original polarity, and are independent of the direction in which the centrifugal force has been operating. Evidently, the factors governing polarity have not been shifted by the centrifuging; they must be inherent in a component of the egg that is not moved by the centrifugal force. Presumably this is the more solid outer layer of the egg, the cortex. Polarity, then, depends on factors localised in the egg cortex; the axial gradients arise only as a result of these factors. There may be some exceptions to this rule. In amphibians, for example, the yolk gradient seems to be of decisive importance for the direction in which the embryo will later develop. If the eggs of an amphibian are turned upside down, and fixed in this position, the heavy yolk material will sink through the cytoplasm from the vegetative side toward the originally animal side of the egg. According to Motomura (1935) and Pasteels (1938-39), this inversion by gravity of the yolk gradient results in an inversion of the polarity of the egg.

Just as, in general, cortical factors are of importance for polarity, so they are for symmetry. Pasteels has made a study of amphibian eggs in which the internal structure was disturbed by gravity or centrifugal force. Their further development proved that the localisation of the primordia of the future embryo, and the direction of its plane of symmetry depend on the interaction of two factors: (1) the yolk gradient, localised in the internal cytoplasm, and (2) a gradient-field in the cortex, the so-called "cortical field". The variable quantity in this field is a hypothetical "C-factor", whose nature is yet unknown. In normal eggs, its maximum intensity coincides with the centre of the grey crescent; its minimum lies at the opposite side. All gradients in this field run therefore in a dorsoventral direction, and the planes of equal intensity intersect the egg surface in a series of concentric circles. This field originates during the formation of the grey crescent.

In sea urchins, cortical factors apparently also play a role. Pease (1939) studied eggs of *Dendraster* which had been centrifuged at a very high speed. He found that the position of the ventral side of the embryo was determined by the interaction of a cortical gradient system with certain substances localized in the internal protoplasm, which accumulate at the centripetal side during centrifuging. This case, therefore, is completely analogous to that of the amphibians discussed above.

While the above-mentioned evidence of the importance of cortical factors for the organisation of the embryo, mainly derived from centrifugation experiments, is of a rather circumstantial nature, more direct evidence has been provided by recent experiments by Curtis (1960-63) with eggs of the toad *Xenopus*. When portions of the cortex from the grey crescent region of uncleaved fertilised eggs were grafted to the opposite side of a second egg, the latter produced two separate embryo primordia. When the grey crescent cortex was excised from fertilised uncleaved eggs, cleavage continued unimpaired, but no embryonic structure developed. Grafts of cortex taken from outside the grey crescent region, placed in the crescent of uncleaved fertilised eggs, led to a splitting of the morphogenetically active region. The grey crescent cortex of the eight-cell embryo still possesses the ability to produce a supernumerary embryo if grafted to younger embryos. But excision of the grey crescent from an eight-cell stage has little or no effect on development, and the embryo of this age has lost the ability to react to cortex grafts. Apart from its effects on the organisation and the determination of the axes of symmetry of the embryo, the cortex is also concerned with cell division mechanisms: grafts of cortex produce an effect on the rate of division of the recipient cell.

These results indicate that the cortex carries a large amount of the "instructions" required to determine where each part of the embryo forms. Different regions of the future embryo are "mapped out", so to speak, in the cortex from an early stage. This may be due to spatial variations in the submicroscopic or molecular structure of the cortex of the egg cell. Such variations could be mainly quantitative and of a continuous nature, in which case the cortical field has the character of a gradient-field. But there might also be qualitative and discontinuous variations in cortical structure, so that the cortical properties are distributed more or less according to a mosaic. Indications for this have been obtained in the eggs of the snail *Limnaea stagnalis*.

In these eggs, treatment with lithium chloride was found to cause characteristic disturbances in development. It results in embryos in which the middle part of the head seems to be reduced so that eyes and tentacles (normally paired) unite on top of the head (so-called cyclopia) (Plate V). In normal development the parts that are suppressed by lithium develop from the cells lying around the animal pole of the egg. When eggs were exposed to a heat shock at an early cleavage stage, on the contrary this middle part of the head seemed to be reduplicated.

From these experiments the conclusion was drawn that

lithium and heat shock treatment act directly on the cortical field, and that this field has the character of a gradient-field, a certain property having its maximum value at the animal pole, and decreasing gradually from this point. The pattern of differentiation was supposed to be dependent on this field, each type of cellular differentiation corresponding to a certain range of values of the field factor. Lithium treatment leads to a weakening of the field, whereas heat shock treatment was assumed to cause a strengthening of the animal gradient-field (Raven, 1958).

Later observations showed, however, that this explanation had to be modified. It appeared that the distortion of the pattern of head organs by lithium or heat shock treatment does not occur in the way predicted by this hypothesis. The primary effects of such a treatment may consist in local deviations of the direction or rate of cell division of single cells. They can hardly be explained on the basis of a mere gradient-field hypothesis, but point rather to very local and discontinuous variations in the cortical field.

Further evidence for the existence of such a cortical mosaic in *Limnaea* was obtained from a study of the cytoplasmic structure of eggs at the very beginning of development, and of their formation in the gonad of the adult. This study indicated that the cortex of the uncleaved fertilised egg indeed carries a mosaic pattern, which is polar, dorsoventral and asymmetric, and that this pattern reflects the mutual positions of the elements surrounding the growing oöcyte in the ovary. It looks, therefore, as if the cortical mosaic becomes "imprinted" upon the egg cell during its growth in the ovary (Raven, 1963).

Experiments on the influence of salts and other substances upon the cortical field in *Limnaea* have shown that its configuration probably depends on local variations in the molecular structure of a system of biocolloids, of which phosphatides form one of the main components, while calcium ions are important for its stability (Raven, 1958). Electron-microscopic observations of centrifuged and lithium-treated eggs have made it probable that the 100 Å thick surface layer of the egg (plasmalemma) carries the field (Elbers, 1959).

Summarising, it appears that the directional organisation of the egg is mainly bound to the egg cortex. The polarity and symmetry of the egg, and thereby the main directions of the future embryo, are determined by the cortical field. This provides the egg with what may be called a system of coordinates, with regard to which all further developmental processes are oriented.

CHAPTER V

Chemodifferentiation

In the foregoing chapters we have reached the conclusion that, at the beginning of development, the egg protoplasm is of a practically homogeneous constitution, and has a very low degree of multiplicity, though there is a polar and bilateral structure, mainly bound to the egg cortex. We shall now see how the spatial multiplicity of the structure increases in the course of development.

Several investigations have made it plain that in many cases the first step in this process consists of a local accumulation of substances that previously were evenly distributed over the egg. This partial separation of the substance mixture of the egg cytoplasm is called *oöplasmic segregation*. The accumulated substances exert a definite influence on the fate of those parts of the germ in which they lie. Originally it was assumed that they supplied the material for certain organs of the future embryo, and for this reason they have been named *"organ-forming substances"*. Later this view proved to be far too schematic. The substances accumulated in certain parts of the germ do exert a determining influence on the further development of these parts; they cannot, however, be regarded simply as building material for the organs concerned. It is better, therefore, to call them *"determining substances"*.

Penners' work (1922-25) on the development of the worm *Tubifex* may be cited as a first example. Soon after fertilisation, accumulations of a special protoplasmic material occur both at the vegetative and at the animal pole of the eggs of this species. These masses are called the *vegetative* and *animal pole-plasm* (Fig. 17 a). The physico-chemical composition of the pole-plasm is slightly different from that of the rest of the protoplasm. For instance, it is rich in certain cell enzymes, bound to mitochondria (Lehmann, 1948; Weber, 1958). During further development, the two masses of pole-plasm unite in the centre of the egg. In *Tubifex*, as in most worms and molluscs, cleavage has a characteristic very regular pattern. First the egg divides into four blastomeres. At the animal side each of these then buds off a number of small cells, the *micromeres*. In this way

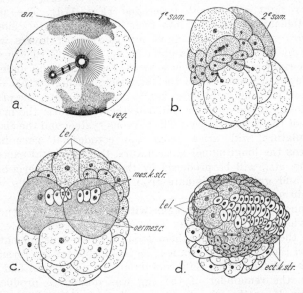

Fig. 17. Development of **Tubifex**. (*a*) uncleaved egg, (*an* = animal, *veg* = vegetative pole-plasm); (*b*) 22-cell stage, first and second somatoblast (*som.*) have been formed; (*c*) formation of the meso-dermal germ bands (*mes.k.str.*) from the primary mesoderm cells (*oermes.c.*), (*tel* = teloblasts); (*d*) older embryo, from the right side, with right ectodermal germ band (*ect.k.str.*), and, at the hind end of the latter, the teloblasts (*tel.*). After Penners.

a number of "micromere quartets" are formed, surrounding the animal pole in a spiral arrangement (so-called "*spiral cleavage*"). At first, the pole-plasm is situated in one of the first four blastomeres. Later, it is distributed over two of the micromeres produced by this cell; these micromeres are larger than the others. They are called the first and second *somatoblasts* (Fig. 17b). In later development, the first somatoblast soon divides into a right and a left half. Each of these halves then divides into a number of large cells, the *teloblasts*. The latter remain at the surface of the germ, and, by budding off small cells toward the anterior end of the embryo, form a left and a right *ectodermal germ-band* (Fig. 17d). The second somatoblast also divides into two. Its two daughter-cells, the *primary mesoderm cells,* move into the deeper layers where, again by budding off cells unilaterally, they give rise to a left and a right *mesodermal germ-band* (Fig. 17c) lying under the ectodermal germ-bands of the same side. During further development, left and right germ-bands gradually unite, the process beginning at the anterior end. Together they form the primordium of the future embryo. The nervous system and the circular musculature will arise from the ectodermal germ-bands, whereas the longitudinal musculature, the segmental excretory organs, etc., will be produced by the mesodermal germ-bands. Skin and gut will originate from the remaining cells which contain no pole-plasm.

Penners irradiated the eggs with a very narrow pencil of ultraviolet light. In this way he was able to kill certain cleavage cells, thereby excluding them from further development. He proved that the pole-plasm was indispensable for the formation of the embryo. If the cells containing the pole-plasm had been killed, the remainder of the egg was unable to produce an embryo. If, on the other hand, a number of other cells, together constituting a large part of the volume, were eliminated, then an abnormally small, but harmoniously built embryo would be produced, so long as the somatoblasts remained intact. Elimination of the first somatoblast led to absence of the ectodermal germ-bands, whereas the mesodermal germ-bands would be present. The killing of the second somatoblast produced the opposite result.

Application of heat causes abnormalities in the cleavage of these eggs. In this case, the pole-plasm is often not restricted to one of the first two cleavage cells only, but distributed evenly

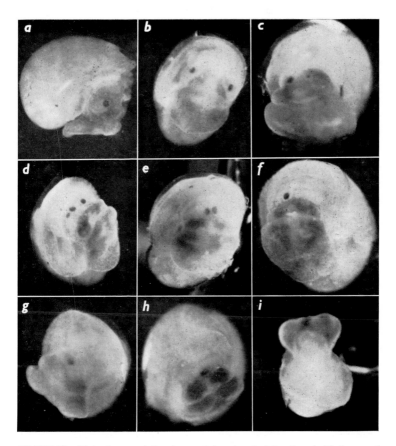

PLATE V. Disturbance of development by chemical treatment. Embryos of a snail, **Limnaea stagnalis**, showing a graded series of malformations following lithium treatment at early stages of development. (a) normal embryo, nine days old, from the right side. (b) The same, front view. (c) The eyes and tentacles have approached each other on top of the head; left eye and tentacle somewhat reduced. (d) Similar embryo, but right eye reduplicated. (e) Eyes contiguous on dorsal side of the head (synophthalmy). (f-h) Eyes fused to single median eye (cyclopy). (i) Embryo viewed from above; eyes and tentacles have disappeared altogether; head region much reduced.

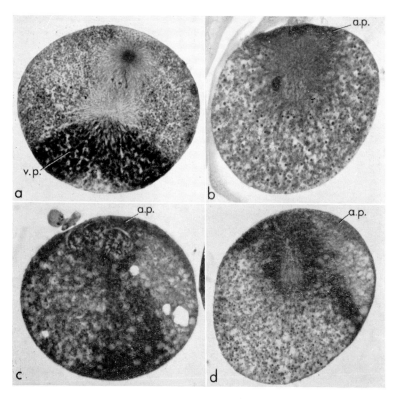

PLATE VI. Oöplasmic segregation and its causes. (a) Freshly-laid egg of a snail, **Limnaea stagnalis,** showing vegetative pole plasm (v.p.). (b) Similar egg at a later stage, showing animal pole plasm (a.p.), which has appeared in the mean time. The vegetative pole plasm substance cannot be distinguished here owing to its different staining. (c-d) Eggs centrifuged at an early stage. Though the egg contents have been stratified (cf. Plate IV) the animal pole plasm (a.p.) has been formed at the right time and place, probably owing to attractions exerted by the egg cortex at the animal side.

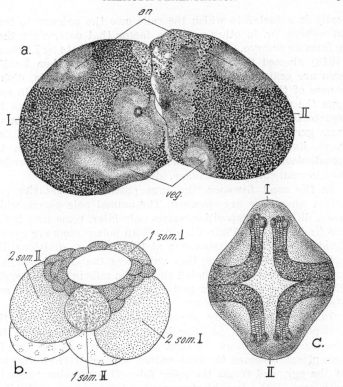

Fig. 18. (a) egg of **Tubifex**, divided into two equivalent cells (I and II). Animal (an) and vegetative (veg) pole-plasm are both halved; (b) a similar egg during cleavage, two first (1.som.I, 1.som.II) and two second somatoblasts (2.som.I, 2.som.II) have been formed; (c) the result is a double monster (so-called **duplicitas cruciata**) with two pairs of germ-bands (I and II). After Penners.

over both of them (Fig. 18a). In further development, each of these blastomeres may form a first and a second somatoblast, containing pole-plasm (Fig. 18b). Two embryo primordia will develop in such a germ; together they form a double monster (Fig. 18c).

These experiments indicate that the presence of pole-plasm

really is a factor to which the cells owe the potency to form an embryo, or, in other words, a factor that determines them to form an embryo. On the other hand, experiments by Lehmann (1948) showed that it is not the only factor. When *Tubifex* eggs are centrifuged or treated with chemical agents, disturbances of the normal pattern of cleavage often occur. In such eggs the substance of the pole-plasms may not pass into the somatoblasts, but come to lie in other cells. Such eggs never form germ-bands. Therefore, the pole-plasm substance only exerts its determining influence when it is situated in the somatoblasts. Apparently other factors, which are dependent on a normal course of cleavage, also play a part.

In the snail *Limnaea,* the eggs possess a vegetative pole plasm when they are spawned. The animal pole plasm, which has a different composition, arises only later, some time before the first cleavage (Plate VI, a-b). Both pole plasms are equally divided among the blastomeres at the first two divisions. But before the next cleavage, they unite near the animal pole, and are then unequally distributed among the cells in the subsequent divisions, the micromeres getting a richer supply of this substance than the macromeres.

In the eggs of many worms and molluscs, the vegetative pole-plasm shows a peculiar behaviour during the first few cleavages. As soon as the first cleavage furrows the egg, the pole-plasm becomes to some extent separated from the rest of the egg, and forms the *polar lobe.* This remains connected with one of the two blastomeres, and later fuses with it so that all the pole-plasm is taken up by one cell. This process may be repeated several times in the subsequent cleavages (Fig. 19). Here again the entire pole-plasm finally finds its way into the first and second somatoblasts. It was shown experimentally that, in these cases too, the pole-plasm is of great importance for further development. Complete removal of the polar lobe results in defective embryos, lacking certain organs; partial removal leads to poor development of these organs (Fig. 20d-f). If parts of the embryo were isolated at later stages, the cells that contained the pole-plasm differentiated into small larvae, which were complete, though not

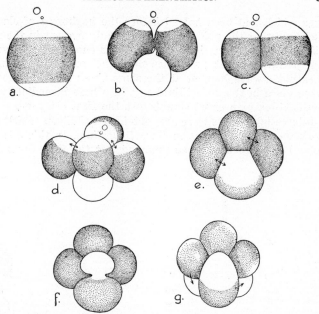

Fig. 19. Cleavage in **Dentalium**. (*a*) uncleaved egg, with animal pole-plasm and polar bodies at the animal pole, and vegetative pole-plasm at the vegetative pole; (*b*) first cleavage, the vegetative pole-plasm, now concentrated in the polar lobe, almost completely separated from the rest of the egg; (*c*) 2-cell stage, the polar lobe has fused with one of the cleavage cells; (*d*) second cleavage, polar lobe has been formed anew; (*e*) 4-cell stage, polar lobe fused with one of the cells; (*f*) third cleavage, polar lobe formed once more; (*g*) 8-cell stage, pole-plasm in one of the cleavage cells. After Wilson.

harmoniously built, showing an overdevelopment of the organs concerned. An isolated group of blastomeres without pole-plasm developed into a defective embryo (Fig. 20a-c). Under abnormal circumstances the polar lobe may be distributed equally over the first two blastomeres; this may result in the formation of a double monster, in the same way as in *Tubifex*. But if the two blastomeres of such eggs become completely separated, each will develop into a complete embryo (Titlebaum, 1928). Wilson (1929) studied the development of fragments of eggs which had been halved before the beginning of cleavage. He found that the pole-plasm was already present in the unfertilised egg. At first, it is distributed homogeneously over the egg plasm; later it assembles near the vegetative pole. This does not always happen at the same time; in some species it takes place in the unfertilised egg, in others immediately after fertilisation, or a short time before the first cleavage.

Interesting results were also obtained by Wilson (1904) in his work on the egg of the mollusc *Dentalium*. In this species, a polar lobe is formed three times, viz. at the first, second, and

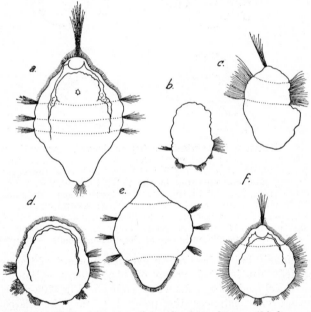

Fig. 20. **Dentalium.** (*a*) normal trochophore larva; (*b*) larva from an isolated blastomere without pole-plasm; (*c*) larva from a cleavage cell with pole-plasm; (*d*) larva from an egg, the first polar lobe of which had been removed; (*e*) as *d*, but only half of the first polar lobe resected; (*f*) larva developing from an egg after resection of the second polar lobe. After Wilson.

third cleavages (Fig. 19). *Dentalium* has a so-called *trochophore* larva, i.e. a larva characterised by a ring of cilia all round the body. This ring separates the anterior *pretrochal* region from the posterior *posttrochal* region. The animal side of the larva bears a group of cilia, the so-called *apical organ*. Removal of the polar lobe at the first cleavage results in a larva without an apical organ, and with a reduced posttrochal region (Fig. 20d). Removal of the polar lobe during the second cleavage produces larvae in which the posttrochal region is still reduced, but which usually possess an apical organ (Fig. 20f). Evidently, the factor responsible for the development of an apical organ is located in the first, but not in the second polar lobe. Therefore, in the interval between first and second cleavages, a shift of the determining substances concerned must take place in the egg protoplasm. This shows that these substances are "preformed" but not "prelocalised" in the egg; they can change their position during development.

The development of the ascidian egg provides another good example of the local accumulation of determining substances. In the protoplasm of newly laid eggs of this group, three different substances can be distinguished. At the animal pole, we find the *ectoplasm*. This has originated in the course of maturation from the nuclear sap of the oöcyte nucleus; on the disappearance of the nuclear membrane, this sap mixes with the surrounding egg protoplasm. A thin layer of *mesoplasm* lies at the surface of the egg, and the rest of it comprises the *endoplasm* (Fig. 21a).

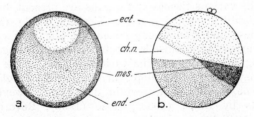

Fig. 21. The egg of an ascidian, before (*a*) and after (*b*) fertilisation. *ect*: ectoplasm; *end*: endoplasm; *mes*: mesoplasm; *ch.n*: chordoneuroplasm. After Conklin.

Soon after fertilisation, complicated streaming movements of these substances can be observed. Ultimately the following distribution is reached: the ectoplasm occupies the animal side, and the endoplasm the vegetative side of the egg; the mesoplasm has accumulated at the ventral side, near the equator, forming the so-called "yellow crescent". Meanwhile a fourth substance, the *chordoneuroplasm* has appeared at the dorsal side of the egg, also near the equator, where it forms the so-called "grey crescent" (Fig. 21b). Consequently, the egg now has a clear-cut bilateral symmetry. In the course of cleavage, the various cytoplasmic substances become localised in different cells. In further development they will give rise to different organs and tissues of the embryo. The skin is produced by the cells containing the ectoplasm; those with the endoplasm will form the gut. The cells of the yellow crescent, which contain the mesoplasm, will supply musculature and connective tissue, and the chordoneuroplasmic cells of the grey crescent will develop partly into notochord and partly into central nervous system.

Several experiments have proved that the cells containing the different types of protoplasm really do have different developmental potencies. Isolated blastomeres produce almost exclusively structures to which they would have given rise in normal development (Conklin, 1905-11). If all the cells containing a particular type of protoplasm are removed, the larva will lack the organs concerned (Von Ubisch, 1939). Centrifugation modifies the relative positions of the protoplasmic substances in the uncleaved egg; such treatment results in embryos in which the various tissues and organs are present, but mixed up in a highly abnormal manner (Conklin, 1931) (Fig. 22).

From these and similar experiments it has been concluded that the various protoplasmic substances may be regarded as determining substances for the various organs of the embryo, i.e. that they steer the development of the cells in which they have become located in a certain direction. However, later experiments by Dalcq (1938), Rose (1939), Reverberi and Minganti (1946-49), and others, point to the fact that the ascidian egg must not be regarded as merely a mosaic of independently developing primordia. Complicated interactions

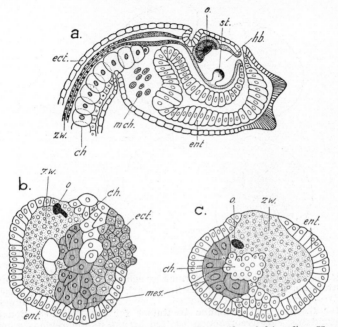

Fig. 22. (a) an ascidian larva, seen from the right, after Kowa-lewsky; (b-c) abnormal embryos formed by centrifuged eggs, after Conklin. *ect* = ectoderm, *ent* = endoderm, *mes* = mesoderm, *mch* = mesenchyme, *zw* = nervous tissue, *ch* = notochord, *hb* = brain vesicle, *st* = statocyst, *o* = eye spot.

between the parts play a role in the development of these eggs too. Yet, apart from these interactions, the localisation of the protoplasmic substances undoubtedly is important for the developmental potencies of the cells. This applies in particular to the mesoplasm, which does apparently determine the differentiation of the cells in which it is present into muscle and connective tissue, and is, therefore, a real determining substance.

Using histochemical methods, Ries (1939) has proved that the mesoplasm of ascidians is characterised by the presence of

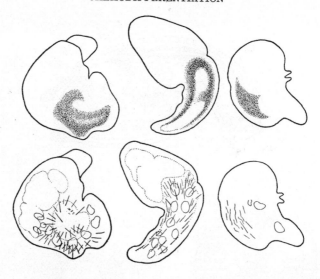

Fig. 23. Abnormal embryos, formed by centrifuged eggs of ascidians. Bottom row: living larvae; the lines mark the places where muscle contractions have been observed. Upper row: the same animals after fixation and benzidine treatment for the demonstration of peroxidases. Positive reaction (dotted areas) in the regions where muscle cells occur. After Ries.

certain enzymes (oxidases and peroxidases), as is the pole-plasm of *Tubifex*. This enzyme-containing protoplasm can be displaced by centrifugation. Now Ries has shown that in embryos developing from centrifuged eggs muscle tissue originates from those cells that happen to harbour the enzyme-containing plasm (Fig. 23).

The oöplasmic segregation, with its local concentration of substances that were at first more homogeneously dispersed, means a transition of a more "probable" to a more "improbable" state of the cytoplasmic system. This implies that there must be forces driving the system in this direction. The fact that these displacements of substances are related to the original polarity and symmetry of the egg, points to the egg cortex

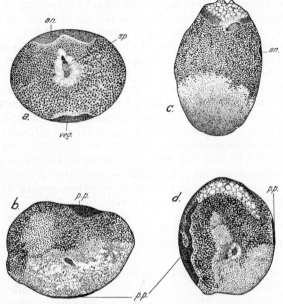

Fig. 24. Eggs of **Tubifex**. (*a*) normal egg, with mitotic spindle (*sp*), and animal (*an*) and vegetative (*veg*) pole plasm; (*b*) centrifuged egg (centrifugal force parallel to the axis of the egg), strong accumulation of pole-plasm (*p.p.*) at the centripetal pole (above), whereas only little pole-plasm is formed at the centrifugal pole; (*c*) egg centrifuged with the centrifugal force acting at right angles to the egg axis, animal pole (*an*) at one side; (*d*) as *c*, somewhat later, centripetal pole above; the pole-plasm has accumulated at the place of the original animal and vegetative poles. After Lehmann.

as the source of these directing forces. This is confirmed by centrifugation experiments. Both in *Tubifex* (Lehmann, 1940) and in *Limnaea* (Raven, 1958) it has been observed that in centrifuged eggs, in which the distribution of the cytoplasmic substances has been profoundly modified, but the cortex presumably is not displaced, the pole plasms assemble beneath the cortex in their normal location and at the proper time

(Fig. 24; Plate VI). This is probably due to attractions exerted by certain parts of the egg cortex on the material concerned.

We have seen above that the original conception of "*organ-forming substances*" supplying the building material for definite organs proved to be an oversimplification. We have then replaced it by the hypothesis of "*determining substances*" which would determine the cells to a certain development. In recent years, however, the view has gained ground that this conception is still too schematic, and that the actual relationships are probably even more complicated. Presumably the differentiation of a cell is not determined by only one determining substance. It is more likely that the direction taken by the development of any cell depends on the mutual quantitative relations of a great number of substances, which together govern the cell metabolism, e.g. enzymes, vitamins, etc. Slight differences in the amount of these substances will modify the "*histochemical equilibrium*" and thereby the direction of development. Peltrera (1940) has given a clear statement of this view on the basis of his work on the distribution of various physico-chemically important substances in the egg of the marine snail *Aplysia*. What we have so far called "determining substances" are probably only, for some fortuitous reason, the more conspicuous components of this complicated system of factors; they may themselves be mixtures of a great number of substances.

We have discussed in this chapter how, after the beginning of development, the originally more or less homogeneous egg cytoplasm acquires a complicated spatial structure because of the concentration of various substances in definite parts of the egg. These substances may at first have been evenly distributed throughout the cytoplasm, their accumulation now taking place under the orienting influence of the egg's "coordinate system", especially its cortex. On the other hand it is possible that some of the substances were not originally present in the egg, but that they arose *de novo* as a consequence of chemical reactions taking place in the egg, e.g. of such an interaction between its axial and cortical gradient systems, as was assumed by Dalcq for the amphibian egg. At any rate, the various parts of the egg, which were all alike before, now begin to show differences in chemical composition. Following J. S. Huxley, we may call this phenomenon *"chemodifferentiation"*.

The differences in physico-chemical composition now existing between the parts of the germ, result in differences in their developmental *potencies*. In broad outline, the egg at first was an "equipotential system"; now, it is divided up into a mosaic of parts with diverging potencies.

Chemodifferentiation may set in at earlier or later stages in development. We have seen above (p. 63) that in ascidians the uncleaved egg is already a mosaic of different cytoplasmic substances. In such a case, the developmental potencies of the different parts of the germ are unequal from a very early stage, and development has the character of a "mosaic development". In other cases chemodifferentiation begins much later, the various parts of the germ remaining equivalent for a long time, both physico-chemically and as regards their potencies. One half of an egg, or two fused eggs, can develop into a single harmoniously built embryo in this case. The one condition for this to occur is, of course, that after the disturbance the cortical field of the egg can adapt itself to the new situation, and readjust itself by the formation of a new equilibrium, essentially similar to the original one. Even in later stages the addition or removal of material may still be regulated in these cases. Such eggs have the character of "regulation eggs". Therefore, the difference between mosaic eggs and regulation eggs is due mainly to an earlier or later occurrence of chemodifferentiation.

During oöplasmic segregation, by the local accumulation of substances under the directing influence of the cortical field, the prodromes of spatial multiplicity carried by the cortex are transferred to the inner cytoplasm, which now becomes different in different parts of the germ. At first, this spatial structure is still very simple, but we can imagine how this situation will be the starting point of increasing complexity. The local concentration of certain substances will start chemical reactions which previously were unable, or almost unable, to take place because of the dilution of the reagents, or because of the presence of inhibiting substances. New substances will thereby be produced in the egg, which again may be locally concentrated. These local accumulations will affect the physical conditions (pH, rH, interfacial tension, electrical properties), and the intensity of metabolism in the areas they occupy. They will also interact, e.g. by mutual attraction or repulsion. All these factors give increasingly complex properties to this physico-chemical system. Once started, chemodifferentiation will steadily proceed, and the egg's intensive multiplicity will be more and more transformed into an extensive multiplicity.

So far we have considered the egg *cytoplasm* only. We shall now discuss a new complication in the system of actions and interactions in the egg, namely the activity of the complex of factors localised in the nucleus.

CHAPTER VI

The realisation of the nuclear factors

We have seen in Chapter II how in fertilisation the nuclei of egg and sperm unite to form a single nucleus, the zygote nucleus or *synkaryon*. This contains the chromosomes of both nuclei, i.e. it has a double set of chromosomes. Genetical and cytological research, which lies outside the scope of this book, has shown that the chromosomes contain a set of material carriers of heritable properties, the so-called *genes*, which are arranged in a linear sequence along the length of the chromosomes. The nucleus of the fertilised egg contains two such sets of genes, one paternal, and one maternal; these transmit the heritable characters of both parents to the new individual. These characters are thus present, in an undeveloped form, in the fertilised egg, and must be realised in the course of the embryo's development. We shall now discuss this process.

During cleavage, the synkaryon divides first into two, then into four, and so on, until a great number of cleavage nuclei has been formed, one in each blastomere.

We have seen above (p. 38) that, according to Weismann's theory, there is a qualitative difference between the products of these nuclear divisions. The hereditary factors (Weismann's "determinants") were assumed to be distributed over the various cleavage nuclei so that, in the end, each cell contains certain of these factors only. We have also discussed the experiments which refute this theory; there are in fact no qualitative differences among the products of the nuclear divisions during cleavage. At least in the earlier stages of development, the cleavage nuclei are identical with each other, and with the zygote nucleus. Each of them contains all the genes, and there are good grounds for the assumption that

71

even in the course of further development only genotypically equivalent nuclei are formed, so that all the nuclei of the organism (possibly with a few exceptions) will contain the full complement of genes typical for the species.

For this reason Weismann's hypothesis that the whole of development is governed by the nuclear factors cannot be correct. On the contrary, nuclei and cytoplasm co-operate in this process. We have discussed the complicated interplay of the developmental processes in the cytoplasm. Evidently, the nuclear factors must in some way or another exert their influence on these processes.

It has been known for a long time that the nucleus is generally indispensable for the life of the cell. Enucleated cell fragments or protists may survive for a while but they show in general no further development, and finally they die. The nucleus is often to be found in those parts of the cell where growth or differentiation takes place. This observation suggests that the nucleus plays a leading, or at least an important, role in these processes. There is good evidence that substances or particles produced by the nucleus pass into the cytoplasm, either migrating through the nuclear membrane, or mixing with the cytoplasm when this membrane has disappeared during nuclear division (cf. what has been said above, p. 63, about the ectoplasm of the ascidians). Especially during oögenesis (the growth of egg cells in the ovary) the nucleus displays a great synthetic activity. Substances produced in the germinal vesicle are transferred to the cytoplasm, where they probably play a part in cytoplasmic growth and in the elaboration of the yolk (cf. Raven, 1961). The most important of these substances is ribonucleic acid. It is synthesised in the nucleus and passed on to the cytoplasm, where it is indispensable for protein synthesis (cf. below, p. 95).

It is well known that the nucleus plays an important role in cell division. It might be said that, at least in uni-nuclear cells, each cell division is preceded by a nuclear division. Cell fragments without a nucleus do not divide any more. This proves that, generally speaking, there is a fixed causal relation between nuclear and cellular division. This statement, however, can be applied to the egg only with certain restrictions. In certain cases, eggs that contain no nucleus at all are nevertheless able to divide. E. B. Harvey (1938), for instance, treated non-nucleated fragments of sea urchin eggs with parthenogenetic agents. This led to cleavage, and even to developmental processes ("parthenogenetic merogony") which went as far as to produce a more or less normal looking blastula, consisting of cells without nuclei, but then came to a stop. The same phenomenon was found in amphibian eggs. This exception to the rule that non-nucleated cells cannot divide can probably be explained as follows. During the growth of the oöcyte and after the disappearance of the nuclear membrane of the germinal vesicle, a large quantity of products of the nucleus has been accumulated in the cytoplasm of those eggs. Divisions can go on in the absence of the nucleus, therefore, until this store is consumed. It is significant, however, that in not even a single case has further differentiation of non-nucleated cells been observed.

The following observation also suggests that there is a certain measure of autonomous activity in the cytoplasm during cleavage. If the polar lobe of certain mollusc eggs is removed at the first cleavage, rhythmic contractions and changes in shape can be observed in the non-nucleated, isolated polar lobe; these movements coincide with the subsequent cleavages of the rest of the egg.

During cleavage the egg cytoplasm is distributed over a great number of cells. In as far as chemodifferentiation has already taken place, there will be differences in physical and chemical properties of the cytoplasm among the various blastomeres. Therefore, the genotypically identical cleavage nuclei will become located in non-identical parts of the egg cytoplasm. The interactions of these nuclei with the cytoplasm surrounding them will result in differences among the reactions that are set going in the various parts of the egg. The nuclear factors will influence the composition of the cytoplasm, and, *vice versa,* the cytoplasm will influence the nuclei and direct the course of their development in divergent ways. We shall first discuss an example of the latter effect.

In the parasitic nematode, *Ascaris megalocephala,* the nucleus

contains a very small number of chromosomes; in one race, the
so-called *univalens* race, there are only two. Consequently, the
division process can very easily be followed in these nuclei, and

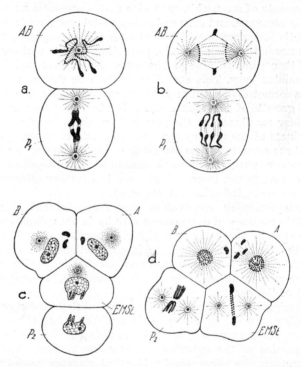

Fig. 25. Chromatin diminution in **Ascaris megalocephala.** (*a-b*)
second cleavage, diminution in *AB*, normal division of the chromosomes in P_1; (*c*) 4-cell stage; (*d*) new arrangement of the cleavage
cells, diminished nuclei in *A* and *B*, diminution taking place now in
EMSt, and normal division in P_2. After Boveri.

this animal constitutes an excellent object for cytological
research. The fertilised egg of the worm contains two long,
curved chromosomes. At the first cleavage, the egg divides into
an animal and a vegetative blastomere. At the second cleavage,

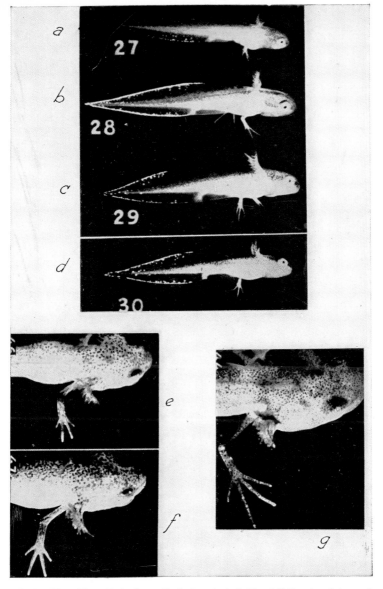

PLATE VII. The effect of genetic factors in hybrids of **Triton taeniatus** and **T. cristatus**. (a-d) four larvae of about the same age. (a) **Triton taeniatus**; (b) **T. cristatus**; (c) **T. cristatus** ♀ × **T. taeniatus** ♂, (d) **T. taeniatus** ♀ × **T. cristatus** ♂. Note the white pigment cells in the caudal fin. (e-g) the shape of the limbs in larvae of the same age. (e) **Triton taeniatus**; (f) **T. taeniatus** ♀ × **T. cristatus** ♂. (g) **T. cristatus**. After Hamburger.

PLATE VIII. The effect of genetic factors on the pigmentation of newts.
(a) larvae of **Triturus rivularis**; (b) larvae of **T. torosus**; (c) hybrids of
T. rivularis × **T. torosus**; (d) transplantation of pigment cell primordium
from **T. torosus** into **T. rivularis**; (e) ditto from **T. rivularis** into **T. torosus**.
The pigmentation in the area occupied by the graft is the same as that in
the species to which the graft belongs. After Twitty.

each of these divides once more into two cells, and, during this division, a peculiar process, known as *chromatin diminution,* takes place in the animal blastomeres. The ends of the chromosomes are thrown off into the cytoplasm, where they disintegrate, and the middle part of each chromosome divides into a number of fragments (Fig. 25a). At this stage, no diminution has yet occurred in the vegetative blastomere. In the next cleavage, however, this will take place in one of its daughter-cells (Fig. 25d). The process is repeated several times, and in the end all cells contain diminished chromatin, except for a single cell, which still has two chromosomes of the original shape. It is from this cell that the primordial germ-cells will develop, which supply the building material for the gonads of the new individual.

Boveri (1910) investigated and experimented on both normal and abnormal eggs of *Ascaris.* He put forward the hypothesis that the nature of the cytoplasm into which the nucleus finds its way at cleavage determines whether or not diminution will occur. The cytoplasm shows quantitative differences according to an axial gradient from the animal to the vegetative pole. The cytoplasm at the vegetative end of this gradient inhibits the occurrence of chromatin diminution in the nuclei. Later observations by Pasteels (1948) have shown that Boveri's view was probably of a too static nature. The cytoplasmic specificity of the primordial germ cells arises only during early development by an unequal distribution of special granules, which are rich in ribonucleic acid, among the cleavage cells. The fact remains, however, that evidently the ultimate development of the nuclei is influenced by local differences in the egg cytoplasm.

The experiments on nuclear transplantation in Amphibia (cf. above, p. 41) give further indications of a cytoplasmic influence on the activity of the nuclei, beginning at a certain stage of development.

The dragonfly *Platycnemis pennipes,* investigated by Seidel (1929-34), provides another clear-cut example of the interaction of nucleus and cytoplasm. This insect has oblong eggs, the hinder ends of which have special properties. Seidel named this region the *activating centre*, because it is of decisive importance for the further development of the egg. If at an early stage of development this centre is killed with a hot needle, or separated from the rest of the egg by means of a tight ligature, no embryo will develop in the anterior part of the egg (Fig. 26 a-d). The activating centre does not begin its activity, however, until one of the cleavage nuclei has reached it. Cleavage in insects differs from that in other animals in that the zygote nucleus, which is located in the middle of the egg, divides repeatedly without accompanying cell-divisions. Therefore, the cleavage nuclei can spread freely throughout the egg. It is only

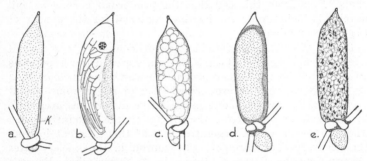

Fig. 26. The activating centre in **Platycnemis pennipes**, and its effect. (*a*) constriction of a very small part of the caudal end of the egg does not prevent the development of the primordium of an embryo (*K*), from which a normal embryo will later develop (*b*), (*c*) constriction of a larger part prevents the formation of an embryo (*d*); (*e*) incomplete ligature prevents the cleavage nuclei (black dots) from reaching the activating centre; again, no embryo is formed in this case. After Seidel.

when a large number of these nuclei has been formed that they migrate to the superficial layer of the egg cytoplasm. Then each nucleus, with the surrounding cytoplasm, is partitioned off from the remainder of the egg to form a separate cell (so-called *"superficial cleavage"*). Now Seidel constricted the eggs just anteriorly to the activating centre so tightly that the cleavage nuclei were unable to reach the centre through the narrow passage that remained. In that case no further development took place (Fig. 26e). But if one nucleus manages to reach the activating centre, a substance is produced there which diffuses into the anterior regions of the egg, and there sets the development of the embryo going. The diffusion of this substance is not impeded by an incomplete constriction of the egg. Evidently, the production of this essential activating substance is due to an interaction between the cytoplasm of the activating centre and a cleavage nucleus. All cleavage nuclei are equivalent also in this respect. Seidel demonstrated this as follows. He killed the hindmost nucleus, a descendant of which would in the normal course of events move into the activating centre, by irradiation with a narrow beam of ultraviolet light. The place of this nucleus was then taken by one of the other cleavage nuclei, and development proceeded in the normal way.

These experiments give us a clue as to the way in which the nuclear factors influence development. During cleavage, the nuclei have become located in areas of different physicochemical properties. At a certain stage, interactions between the nuclei and the cytoplasm occur. They vary in different parts of the egg because of the cytoplasmic differences. Nuclear factors, which have so far remained inactive, can begin to unfold their activity once the nuclei come to lie in a suitable environment. It may be said, therefore, that they are activated by the surrounding cytoplasm. In other parts of the egg, where the cytoplasm is of a different composition, the reaction in question will not occur, but other nuclear factors may be activated there.

We have not yet entered into the question of the nature of these nuclear factors. It is natural in this connection to think of the genes, which we may regard as localised in the chromosomes. An experiment by Boveri (1907) demonstrates that the chromosomes do indeed play an important role in development. If sea urchin eggs are fertilised with concentrated semen, into many eggs two sperms will penetrate simultaneously. As a rule both sperm nuclei fuse with the egg nucleus in such eggs, forming a "triploid" zygote nucleus, containing three sets of chromosomes. At the first cleavage the majority of such abnormal eggs divides at once into three or four daughter-cells simultaneously. The chromosomes of the triploid nucleus are distributed at random over the three or four spindles, so that the number of chromosomes arriving at each of the poles and combining to form a daughter nucleus varies widely in individual cells.

By two different methods Boveri made a careful study of the further development of such eggs. Sometimes he allowed complete eggs to develop, but at other times he separated the three or four simultaneously formed daughter-cells. Boveri found that, on the whole, the isolated cells developed poorly, but that there were great individual differences. Some cells died very soon, others developed better and produced fairly normal small larvae. Experiments with whole eggs had the same result. As a rule, cleavage was normal, but then development came to a stop in certain sectors of the egg, and the cells of those parts died. In general, eggs which had divided into three developed better than those which had divided according to the tetraster type.

However, it appeared that the success of development did not depend only on the *number* of chromosomes present in the nucleus. Sometimes parts with a larger number of chromosomes

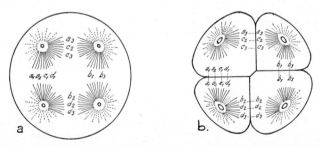

Fig. 27. Diagram of the distribution of the chromosomes in a tetraster division, taking place in a sea urchin egg after double fertilisation, taking 4 as the haploid chromosome number. (*a*) the 12 chromosomes of the triploid egg become distributed at random over the 4 spindles; (*b*) consequently, only one of the cells (bottom left) receives a complete set of chromosomes, a, b, c and d. After Boveri.

were less successful than other parts with a smaller chromosome number. Boveri concluded that the result of development is determined by the *combination* of chromosomes. He suggested that there were qualitative differences between the chromosomes, and that only those parts of the egg could develop normally which had at least one complete (haploid) set of chromosomes in their nucleus (Fig. 27).

Later investigations have shown, indeed, that absence of one chromosome, or even of a small part of a chromosome, is sufficient to bring development to a stop at an early stage. Poulson (1940-45), for instance, studied the consequences of "deficiencies" of varying extent in the X-chromosome of the fruit fly, *Drosophila*. Complete absence of this chromosome resulted in disturbances as early as the end of cleavage, at the moment when the cleavage nuclei migrate to the surface of the egg (cf. p. 76). In the case of absence of one particular half of the X-chromosome the disturbances also became manifest at the same stage, but the distribution of the cleavage nuclei over the surface layer was slightly less abnormal. If the other half of the chromosome was absent instead, the superficial cell layer developed normally, but no embryo was formed. Finally, absence of small parts of the chromosome led to certain characteristic aberrations at later stages, e.g. in the development of the nervous system or of the gut.

These investigations prove that the full complement of chromosomes is necessary for normal development. Since the absence of very small parts of a chromosome may result in the disturbance of very definite developmental processes, it seems probable that the normal course of these processes depends on the presence of special genes, located in these parts.

We owe our insight into the way in which the genes work in the first place to a number of investigations on so-called *lethal factors,* i.e. genetic factors which lead to the death of the embryo at some given moment during its development. They arise by mutation, and show a normal mendelian inheritance. Partly they are due to the loss of chromosomal material (deficiences) or other structural changes in the chromosomes; in other cases, however, they concern mutations of single genes.

Each lethal factor was found to exert its influence at one definite stage of development, and often also at one definite place, by directing the course of development of some organs or parts of the embryo into abnormal channels, leading to the death of the embryo. In many cases, however, one gene proved to influence several developmental processes, localised in different parts of the germ, at the same time. Side by side with the dying tissues, other organs on which the lethal factor has no visible influence often occur in the same embryo. If such tissues are removed in good time, and grafted into a healthy embryo, they are able to develop normally in spite of the fact that the nucleus of each of their cells contains the lethal factor that caused the death of the other tissues. The importance of work of this kind lies in the fact that from the mode of operation of these lethal factors, the conclusion can be drawn that the corresponding processes in normal development are governed by normal genes. For the lethal factors have originated by the loss or mutation of normal genes. Consequently, the resulting disturbance in development must be due to the disappearance, or modification, of the normal action of these genes. Indirectly, therefore, these observations give us information as to the time and place at which the genes exert their influence in normal development.

The genes co-operate with the cytoplasmic factors. We cannot find out, simply by looking at the result, what has been the part played by each in the origin of the embryo. In order to assess the relative importance of cytoplasmic and nuclear factors, we must combine them in such a way that the groups can be told apart by their specific recognition marks. This can be done by combining nuclear material of one species with egg cytoplasm of another; the properties of the embryo that results from such a combination may give some information on the problem under consideration.

The crossing of two related species is the simplest way to achieve this. It results in eggs, the cytoplasm of which is almost entirely maternal, whereas the nucleus contains genes of two species, because it arises from the fusion of egg and sperm nuclei. One might expect that it would be very easy to conclude, from the development of such hybrids, when and how the action of the paternal genes becomes manifest in development.

In practice, however, this type of crossing experiment is fraught with certain difficulties which seriously reduce its usefulness for the solution of our problem. Firstly, crosses of somewhat distantly related species do not as a rule give viable embryos. Broadly speaking, it can be said that the more remote the two parent species are from each other, the earlier the development of the hybrids will come to a stop. However, taxonomic affinity is not the only factor to be considered, as may be seen from the fact that sometimes the result of the combination of eggs of species A with sperm of species B is entirely different from that of the opposite combination. The arrest of development may be due to different factors. Sometimes disturbances occur in the nuclear divisions so that the distribution of the chromosomes among the cells is abnormal. In other cases, the cells show pathological phenomena from a given moment onwards. These may be restricted to certain parts or organs of the germ, whereas other parts are not affected, or, on the other hand, they may be present more or less equally in all cells. This in itself does prove that for normal development nucleus and cytoplasm must be "attuned" to each other, but it makes a further analysis of the role of each group of factors impossible in these cases. Hybridisation of more closely related species is attended with another difficulty. The characters in which these species differ do not appear until a very late stage in development. During the major part of the process, the hybrid is indistinguishable from either parent, and if there are any influences of the paternal genes, they cannot become visible. Their activity becomes manifest only at the end of development, when all the more essential steps in the process have long ago been completed.

In accordance with the foregoing remarks, the result of a cross between two sea urchin species depends very much on the species used. In the most favourable cases, the hybrids will develop into normal pluteus larvae. This will occur, e.g., in the case of the fertilisation of eggs of *Sphaerechinus granularis* with sperm of either *Psammechinus microtuberculatus* or *Para-*

centrotus lividus. The plutei of these species (**Fig. 28**) show marked differences. Those of *Sphaerechinus* are of a plumper shape; the relative lengths of the arms are different, and the structure of the calcareous rods is more complicated. One of the latter, the anal rod, consists of 3-5 parallel rods with cross connections. Another, the apical rod, has three prongs at its top. In *Paracentrotus*, the anal rod is single, and the apical rod club-shaped, but not branched (in *Psammechinus* the latter shows an incipient branching; see below, p. 87). Now it was found that the hybrid larvae were intermediate, both in shape and in structure of the skeleton. Their anal rods consisted of

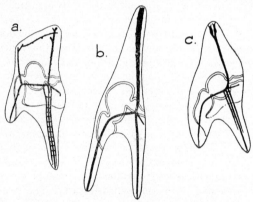

Fig. 28. Pluteus larvae of (*a*) **Sphaerechinus granularis**, (*b*) **Psamm-echinus microtuberculatus**, and (*c*) the hybrid **Sphaerech. gran.** × **Psammech. microtuberc.** After Boveri.

2-3 rods, without cross connections; their apical rods were club-shaped, but also showed a tendency to branch (Boveri, 1889-95). It is a remarkable fact that, apart from intermediate larvae, other larvae are sometimes found with a preponderance of paternal characters (*"patrocline"* larvae). Conditions in the culture seem to play a role here; in particular the temperature appears to influence the structure of the calcareous skeleton.

In this case, the influence of the paternal chromosomes on development was unmistakable. Other crosses, however, had

different results. In many cases, the hybrid was entirely similar
to the species to which the eggs belonged (so-called *matrocline
hybrids*), e.g. in the cross *Sphaerechinus* ♀ × *Arbacia* ♂. In
other cases, the incompatibility of the paternal chromosomes
with the surrounding "foreign" cytoplasm manifested itself in
a somewhat dramatic way. We have already seen that the
cross *Sphaerechinus* × *Psammechinus* (or *Paracentrotus*) gives
normal hybrids of intermediate structure. The reciprocal cross,
however, *Psammechinus* (*Paracentrotus*) × *Sphaerechinus* has
an entirely different result. The paternal chromosomes do unite
with the egg nucleus, but in the first or second division, 16 of
the 20 *Sphaerechinus* chromosomes are eliminated, and such
development as takes place is matrocline.

In amphibians, too, the result of crossing experiments
depends very much upon the nature of the species crossed.
Good development of the hybrids occurs, e.g., in the cross
Bufo communis ♂ × *B. viridis* ♂, whereas in the reciprocal
cross the hybrids develop poorly, and in the cross *Rana esculenta* × *R. fusca* development ceases already at the gastrula
stage. In other crosses one obtains so-called "false hybrids",
in which the penetrating sperm has set development going,
but does not otherwise take part in the process (e.g. in *Rana
esculenta* × *Bufo viridis*).

J. A. Moore (1941) studied the development of a great
number of American frog hybrids. In many cases, these developed normally until the early gastrula stage, but then development ceased. In other crosses development proceeded further.
In all cases, the rate of development was entirely similar to
that of the species which had produced the eggs, up to the
beginning of gastrulation. In those hybrids which developed
beyond this stage, the influence of the paternal chromosomes
on the rate of development became noticeable during or after
the completion of the gastrula stage.

Further it has been found in frog hybrids which stopped
developing at the gastrula stage, that simultaneous deviations
occurred in respiration and carbohydrate metabolism (Barth
and Jaeger), and the synthesis of ribonucleic acid which normally begins at this stage did not set in (Brachet, 1952).

Evidently the presence of the foreign chromosomes inhibits certain processes by which the nucleus influences the cell metabolism (cf. p. 72). Curiously enough, parts of such "lethal hybrids", grafted into normal embryos, proved able to continue their development more or less normally. This is accompanied by a distinct synthesis of ribonucleic acid in the grafts. This "revitalisation" of the lethal tissues must probably be ascribed to the fact that the substances indispensable to metabolism, which the cells of the hybrid cannot themselves synthesise, are now supplied by diffusion from the environment.

The influence of the paternal nuclear factors on further development has been carefully studied in some newt crosses. Hamburger (1936) crossed *Triton taeniatus* and *T. cristatus*. The hybrid larvae initially were the same size as those of the maternal species. Later, however, differences developed. *T. cristatus* grows more rapidly than does *taeniatus*. The hybrids are intermediate, though *cristatus* × *taeniatus* grows somewhat faster than the reciprocal hybrid. This difference between the hybrids shows that growth does not depend only on the nuclear factors, but on the cytoplasmic factors as well: the composition of the nuclei is identical in the two hybrids, but their cytoplasm is different, being almost exclusively maternal. Another difference between the parent species is that the white pigment cells on the caudal fin appear earlier in *Triton cristatus* than in *taeniatus*, and have somewhat different patterns in the two species. In both hybrids, the appearance of this character coincides with that in *cristatus*, and at that time the pattern, too, essentially tallies with that of *cristatus* (Pl. VII a-d).[1] Later, however, it begins to approach the *taeniatus* pattern. The similarity in behaviour of the two hybrids on this point shows that cytoplasmic factors do not play a role here, but that the formation of these pigment cells is governed entirely by the nuclear factors. Finally, there are marked differences between *taeniatus* and *cristatus* larvae in the structure of the fore-limb, viz. in shape and length of the fingers. The limbs of the hybrids are intermediate between those of the parent species (Pl. VII

---

[1] Facing page 74.

e-g). There is no difference between reciprocal hybrids on this point, so that here again only nuclear factors seem to play a role.

In a study of American species of the genus *Triturus,* Twitty (1936) was able to carry the analysis still further by means of a combination of transplantations and hybridisations. There are characteristic differences in pigmentation between the species studied by him. In *Triturus torosus,* the black pigment cells form a well defined black band on each side along the base of the dorsal fin. In *T. rivularis,* they are dispersed all over the sides of the body. The pigmentation of hybrids between these species is more or less intermediate between that of their parents. Evidently the distribution of the pigment cells over the sides is governed by the nuclear factors (Pl. VIII a-c).

Twitty tried to find out how the nuclear factors influence the distribution of the pigment cells. These cells originate dorsally, from the embryonic neural folds (cf. p. 104), and spread from there over the sides of the body. In *T. torosus,* they accumulate in the dorsal regions of the sides, above the muscle segments, whereas in *T. rivularis* they become evenly scattered over the whole flank. The differences in pigmentation, therefore, are due to differences in the migrations of the pigment cells. These in their turn might depend either upon differences in the nature of these cells, or upon differences in the orienting and attracting influences from the environment to which they are exposed. By means of transplantation experiments, Twitty and his collaborators proved, indeed, that both the skin epithelium and the muscle segments of various species have different properties in this respect. But the specific differences in the pattern of pigmentation proved to be due mainly to the properties of the pigment cells themselves. In *T. torosus,* the latter are strongly attracted by the muscle segments, and to a lesser degree by the central nervous system. The pigment cells of *T. rivularis* lack this property; hybrids possess it to a smaller extent. Parts of the pigment cell primordia in the neural fold of *torosus* embryos were replaced by corresponding parts of *rivularis* embryos, and *vice versa.* In this case, the pigment cells of the *torosus* graft accumulated over the muscle segments of the *rivularis* host, whereas the pigment cells of the *rivularis* graft scattered evenly over the side of the *torosus* host. In other words, the graft's own specific pattern will develop even in the foreign host (Pl. VIII d-e).[1] Twitty (1945) has shown in later experiments that the dorsal accumulation of the *torosus* melanophores is a secondary phenomenon. As soon as they have reached a certain degree of differentiation, the cells withdraw their processes, thereby moving closer together. In *T. rivularis*, the melanophores always remain at a lower level of differentiation; they do not become so strongly pigmented, and do not show secondary accumulation. The nuclear factors, therefore, are directly responsible for a difference in the ultimate level of differentiation of cells of this category; indirectly, this results in a modification of the pattern of skin pigmentation.

In the cells of hybrids, the paternal nuclear material is always combined with the maternal chromatin of the egg nucleus. This means that the influence of the paternal genes on the foreign cytoplasm will not be able to express itself unhampered. It is always more or less masked by that of the maternal genes. If a better picture of the mutual influence of nuclear and cytoplasmic factors is desired, a nucleus of species A should be combined with cytoplasm of species B in the absence of nuclear material from the latter. If this can be achieved, the foreign nucleus will not be able to use the "mediation" of this material when influencing the cytoplasm, but the two must interact directly. Properties of the hybrid which are characteristic of A are then undoubtedly caused by the nucleus, and B-characters equally certainly by the cytoplasmic factors. Such combinations can indeed be realised in a number of cases. Non-nucleated egg-fragments or whole eggs, the nucleus of which has been removed or killed, can under certain circumstances develop after fertilisation with nuclear material derived only from the sperm nucleus present in the cell. The general term for this phenomenon is *merogony*. If combined with hybrid fertilisation — the sperm nucleus belonging to another species — it is called *heterospermic* or *hybrid merogony*. The ideal case described

---

[1] Facing page 75.

above is realised here. The heterospermic merogone consists of cytoplasm of one species, and a nucleus of another.

In experiments on heterospermic merogony, the same difficulties are generally met with as in normal hybridisation: hybrid merogones in which distantly related species are combined, are of low viability, and as a rule they soon die. It is interesting to note that, just as in the case of lethal factors (p. 80), this "lethality" often expresses itself in the disturbance of definite developmental processes at a definite stage, whereas other processes, in other organs, may take a fairly normal course. The combination of more closely related species gives better results, which however are not quite so interesting from our point of view, because the differences between the parents do not become manifest till late in development. Yet these experiments furnished some results, in particular again in sea urchins and amphibians.

Hörstadius (1936) used the following method for the production of sea urchin merogones. With a sharp needle he removed from the egg a small segment containing the nucleus. The enucleated eggs were then fertilised. In this way he obtained hybrid merogones between *Psammechinus microtuberculatus* and *Paracentrotus lividus*. These developed well, and produced harmoniously built plutei. It is true that the plutei of the parent species are very similar to each other, but after a very minute investigation, Hörstadius was able to demonstrate, by means of slight differences found in the shape of the calcareous rods, that the sperm nucleus had a clear-cut influence on these organs. The tips of the apical rods are slightly more irregular in *Psammechinus* than in *Paracentrotus,* also they show a slight tendency to branch. In this respect, the hybrid merogones behaved in the same way as the species that supplied the sperm nucleus (Fig. 29). Evidently this character is governed by the nuclear factors. Some further information on their mode of operation may be obtained from the combination of this result with observations by von Ubisch (1937) on the way in which the calcareous skeleton of the pluteus develops. It is built by mesenchyme cells of the gastrula, which are grouped to the left and right of the archenteron, and fuse into a so-called

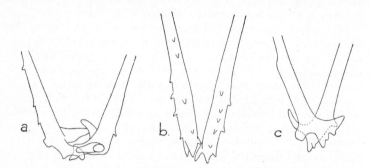

Fig. 29. Apical rods of the skeleton of the pluteus larva of (*a*) **Psammechinus microtuberculatus**, (*b*) **Paracentrotus lividus**, and (*c*) the hybrid merogone **Parac. liv.** (cytoplasm) × **Psammech. microtub.** (nucleus). In the latter, the shape of the apical rods is more similar to that of **Psammechinus**. After Hörstadius.

syncytium. The material of the rods is then secreted within this syncytium (Fig. 31). The viscosity of the cytoplasm of these cells appears to be an important factor in the final determination of the shape of the calcareous rods. Lower viscosity, which may occur as a specific character in some species, and may be caused by a rise in temperature in others, leads to more complicated branching of the skeleton. Therefore, the primary influence of the nuclear factors in the merogones mentioned above might consist in a modification of the viscosity of the protoplasm of the skeleton building cells.

Interesting results were also obtained by von Ubisch (1954) with heterospermic merogones of *Sphaerechinus* (cytoplasm) and *Paracentrotus* or *Psammechinus* (nucleus). Most of these merogones cease development at the blastula or early gastrula stage, but very few (16 out of 3000) reached the pluteus stage. The skeleton of these plutei is entirely paternal. Their external shape is at first paternal, but with increasing age the maternal shape becomes more and more predominant. Apparently, the growth rate of the ectodermal layer is determined by maternal, hence cytoplasmic factors.

In Amphibia, merogonic development may be brought about by several means. G. and P. Hertwig killed the nucleus of the egg by irradiation with mesothorium, and then fertilised with normal sperms. Or, on the other hand, the nucleus may be removed from the egg with a fine pipette. Homospermic merogones, i.e. merogones in which the sperm nucleus belongs to the same species as the egg cytoplasm, are often able to differentiate into a harmoniously built embryo. In heterospermic merogones between newts of the genus *Triton,* development will cease at a given stage, and this will be earlier, the more distant the parent species. In any case, development will stop before the stage at which the differential characters between the parent species begin to show (Baltzer, 1920). For that reason, these merogones themselves, again, do not help us much towards the solution of our problem. Hadorn (1932) found, however, that e.g. in merogones of *Triton palmatus* (cytoplasm) and *T. cristatus* (nucleus) not all tissues are damaged to the same extent, although the merogones die at an early stage. The mesenchyme cells of the head die, and the muscles and nervous tissue are also incapable of differentiation, but skin ectoderm, heart, notochord, gut, and pronephros, are apparently unimpaired. This shows that the antagonism between nucleus and cytoplasm of different species is not manifested equally strongly in all parts, but especially where the cells must achieve certain differentiations. This is another indication that the "activation" of the nuclear factors takes place at different times in the different parts of the embryo. In related species of *Triton* these nuclear factors may be partly identical, and partly different. Normal differentiation will be possible in those cells in which no nuclear factors are activated during differentiation, or only such factors as occur in the maternal species too, and which therefore "fit in" with the cytoplasm. Disturbances, sometimes resulting in the death of the cells, will only occur where factors are activated which differ in the two species. Hadorn managed to keep the unimpaired tissues of the merogones alive to much later stages by grafting them into normal embryos. There they differentiated normally. In a few cases, skin ectoderm of such a merogone, when grafted into an embryo of *Triton alpestris,* remained intact until after the metamorphosis of the host. Meanwhile, the specific characters of this tissue had become visible, and they were found to agree entirely with those of the species which had supplied the cytoplasm, i.e. *Triton palmatus*. No specific influence of the nucleus of *T. cristatus* was observed.

Comparable experiments were made by Dalton (1946) with American newts of the species *Triturus torosus, T. similans,* and *T. rivularis,* the same species that were used by Twitty (see p. 85). Here, too, the hybrid merogones die early, before the development of specific differences in pigmentation. But it was again possible to keep the primordia of the pigment cells alive for a longer time by early transplantation into a normal embryo. We have already seen that the pattern of pigmentation depends upon the properties of the pigment cells. Dalton found that it is determined mainly by the nuclear factors, though the cytoplasm has some influence.

Finally, mention must be made of an experiment by Baltzer (1947) on the pigmentation of merogonic hybrids between black and white axolotls. The difference in pigmentation between these races is caused by a single gene, black being dominant over white. The character does not depend on the pigment cells themselves, but on the epidermis covering them. In the black axolotl, this is assumed to secrete a substance, probably an oxidase, which is necessary for the production of melanin in the pigment cells. In the white axolotl, the production of this substance is reduced (Dushane, 1935). It appeared from Baltzer's experiments that merogones consisting of cytoplasm of the black axolotl, and of nuclear material of the white race, had the same type of pigmentation as the white race. Even after early transplantation into a black larva, skin grafts of such a merogone developed a pigmentation of the white type. In this case, the nature of the epidermis, on which the pigmentation depends, appears to be determined entirely by the nuclear factors.

The above-mentioned experiments all have led to the view that in general the intervention of the nuclear genes in development takes place in such a way that a certain gene becomes "activated" at certain moments and in special cells in response to the composition and momentary state of its cytoplasmic environment. A visible expression of such "activation" has recently been found in the giant chromosomes in certain organs of the larvae of various dipterous insects. Such giant chromosomes consist of alternating transverse light and dark bands in a regular pattern, which is supposed to reflect the sequence of the genes along the length of the chromosomes. Occasionally, one of the dark bands (chromomeres) becomes swollen and inflated, and its chemical composition is changed (for instance, ribonucleic acid appears in it). The localisation of such "puffs" varies in different organs; moreover, a certain "puff" has only a temporary existence, its waxing and waning being correlated with definite phases of development. "Puffing" in distinct loci can be induced by external factors, e.g. by the injection of a moulting hormone in the larva (Beermann, 1963). The most obvious explanation of these observations is that the "puffing" phenomenon is a cytologic expression of the reversible activation of certain genes.

The activation of certain genes in the nucleus as a rule results in the production of substances in the cell which in their turn influence other developmental processes. Such substances produced as a consequence of gene activity may be restricted to the cells in which they have been formed, and may influence the chemical, physical, morphological, or physiological properties of these cells only. Probably the pigment cells of newts (see p. 86) and the skeleton-building cells of sea urchins (see p. 88) are examples of this kind. In other cases, however, the substances produced under the influence of certain genes may diffuse from the cells in which they are synthesised, and influence the determination of other parts of the embryo as well. Their influence in that case may either be restricted to immediately adjacent cells (embryonic induction, cf. below, p. 121) or, when they are transported, e.g. with the blood, throughout the body, even the development of remote parts of the body may be influenced by them. Examples of this latter kind have been studied in particular in insects (Plagge, 1939; Ephrussi, 1942).

In the fruit fly *Drosophila melanogaster* two mutations occur, known as "vermilion" (*v*), and "cinnabar" (*cn*), respectively. The eye colour of these mutants is lighter than that of normal individuals. The eyes of the latter are coloured dark red by the combination of a red and a brown pigment. The brown pigment is absent both in *v* and in *cn* mutants. Both mutants are due to single gene mutations. If grafted early into a normal animal, the eye of a *v*-mutant developed the same dark red colour as is found in the wild form. The same was seen on transplantation of *v*-eyes into a *cn*-animal. Eyes of a *cn*-larva also became dark red when transplanted into a normal animal. However, they remained light if grafted into a *v*-animal. The result did not depend upon which particular part of the body the eyes were grafted into. It can be concluded from these observations that the whole body, both of the normal animal, and of the *cn*-mutant, contains a factor, probably a substance, which gives rise to brown pigment in the eyes of the *v*-mutant, and which is evidently absent in the *v*-mutant itself. Similarly, a factor necessary for the formation of brown pigment in *cn*-eyes is present in the wild form, but not in either *v*- or *cn*-mutants. Apparently the cells of certain organs (eyes, Malpighian tubes, fat-body) of the normal animal produce an excess of these substances. Transplantation of these organs from a normal animal into a *v*- or *cn*-mutant resulted in normal, wild type coloration of the host's eyes. Injection of lymph from a normal animal had the same effect. Therefore, the substances concerned must be present in lymph as well. Further experiments showed that in a normal, wild type *Drosophila* a reaction-chain occurs which first leads to the production, under the influence of a single gene, of a so-called *v+* substance. Under the influence of a second gene this substance is transformed into a second substance, the *cn +* substance, which in its turn is indispensable for the normal pigmentation of the eyes. In the mutants, this catenary reaction is interrupted in various places. In "vermilion", the gene that takes care of the production of the *v +* substance has fallen out. In "cinnabar", the *v +* substance is produced, but it cannot be transformed into the *cn +* substance because of the absence of

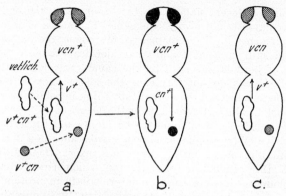

Fig. 30. **Drosophila.** (*a*) the fat body of a normal animal ($v+cn+$) is transplanted into a $vcn+$ animal, where it secretes $v+$ substance; (*b*) the host tissues transform this into $cn+$ substance, so that both the eyes of the host, and a grafted $v+cn$ eye can now develop normal pigmentation; (*c*) the same transplantation into a vcn animal has no effect, because there the $v+$ substance cannot be transformed into $cn+$ substance. After Ephrussi and Chevais.

the necessary gene (Fig. 30). As we have seen, these substances are produced by certain cells of the body under the influence of certain genes, and then spread throughout the body via blood or lymph.

A single gene substance may influence several different developmental processes. In the moth *Ephestia,* one substance influences the speed of development, and the viability of the embryo, the pigmentation of eyes and skin of the caterpillar, and the pigmentation of eyes, brain, and sex organs of the imago. Another remarkable discovery is that gene substances are not specific. The $v+$ and $cn+$ substances of *Drosophila melanogaster* were also found in other flies, parasitic wasps, and butterflies. The catenary reaction found in *Drosophila* apparently occurs in insects in general, and the existence of mutations entirely analogous to "vermilion" and "cinnabar" in *Drosophila* has been demonstrated in *Ephestia* and in the wasp *Habrobracon,* where, as a matter of fact, they had been known for a long time under different names. On the other hand, the end point of the reaction is not always the same. There is a wide variety of eye pigments in different insects. Moreover, the pigments of eyes and skin in *Ephestia* are not identical, though the production of both is governed by the same gene. Evidently, the gene substances do not influence the last stages of pigment formation, but instead they constitute a link in earlier processes. Later work has thrown some light on the nature of these processes. The $v+$ substance was found to be identical with kynurenin, a derivative of tryptophane. The $cn+$ substance is a derivative of kynurenin. Apparently, the genes exert their influence through the production of enzymes which first transform tryptophane — itself a common intermediate in protein metabolism — into kynurenin, and subsequently kynurenin into other substances which are the building stones for skin and eye pigments.

Investigations of recent years have considerably improved our insight into the way in which genes work (cf. Raven, 1961). They have made it highly probable that in general each gene controls the synthesis of one enzyme or other protein. They further indicate that deoxyribonucleic acid (DNA) in the chromosomes forms the essential component of the genes and the carrier of the genotypical information. This substance forms long macro-molecular chains, the axis of which is formed by alternating phosphate and sugar (deoxyribose) molecules, while the latter each carry a nucleotide base as a side-chain. These nucleotide bases belong to four different kinds, the purines adenine and guanine, and the pyrimidines thymine and cytosine. Each macromolecule of DNA carries many thousands of these bases. The sequence of the four kinds of nucleotide bases along the chain determines the specificity of the gene, and represents the genotypical code.

In the natural state two of these chains are wound around each other, and connected by transverse bridges consisting of paired adenine-thymine or guanine-cytosine groups, linked together by hydrogen bonds. In the replication of the DNA molecule the two complementary halves separate, and each half is completed by attracting the matching nucleotides in the right order, so that two new molecules are formed, which are identical with the parent molecule.

Proteins also consist of long chains, formed in this case by sequences of amino acids. Twenty different amino acids generally take part in their formation. Just as the sequence of the nucleotides determines the specificity of the DNA molecule, so the sequence of the amino acids is supposed to be responsible for the nature and properties of the protein.

Now it appears that the sequence of the four kinds of bases along a nucleic acid chain controls the sequence of the amino acids along a protein chain. More specifically, it is thought that the position of each amino acid in the protein is determined by a certain combination of three successive nucleotide bases in the DNA. There are indications that this nucleic acid code is identical for all organisms, and therefore has a very fundamental biological importance.

In order for a gene to control the synthesis of a certain protein, it is necessary that the information which it contains is transported to the place in the cell where protein synthesis actually occurs. As a rule this is not the nucleus but the cytoplasm. It is here that ribonucleic acid (RNA) comes into the picture. In its molecular structure it closely resembles DNA, deoxyribose being replaced by ribose and thymine by uracil. It appears that most, if not all, of the RNA of the cytoplasm is synthesised in the nucleus. Three kinds of RNA, with different functions, can be distinguished: (a) ribosomal RNA, (b) transfer RNA, and (c) messenger RNA. Ribosomal RNA is found in the ribosomes (cf. above, p. 37) which are the main sites of protein synthesis of the cytoplasm. Transfer RNA (or "soluble" RNA), consisting of relatively small molecules, plays a part in the transport of "activated" amino acid molecules to the ribosomes. Finally, messenger RNA is probably formed along the DNA strands of the chromosomes, of which it forms a copy. It then passes from the nucleus to the ribosomes where it forms (or contributes somehow to the formation of) a template along which the amino acids are assembled in the right order, and linked together to form a polypeptide chain.

It is therefore the messenger RNA which carries the geno-typical information from the nucleus to the cytoplasm.

It is evident that these new insights in "molecular biology" are of great importance for our understanding of the role played by the genes in development.

With respect to the problem of the control of gene activity during embryonic development, recent work in different fields of biology has suggested various possibilities of the way in which such control could operate (cf. Waddington, 1962). It seems too early, however, to go more deeply into this problem here.

CHAPTER VII

The topogenesis of the embryo

In the foregoing chapters we have seen how a series of processes transforms the initially fairly homogeneous structure of the egg into a system with a considerable degree of spatial multiplicity. First, the primary co-ordinate system of the egg, expressed in its polarity and symmetry, and carried by the cortical field, gave rise to the localisation of determining substances. This was followed by the production and local accumulation of new components of the cytoplasm, resulting from the chemical and physical processes now started in the egg. Differences began to occur, both chemically and physically, among the various parts of the egg. Finally, reactions between nuclei and cytoplasm took place as a consequence of the activation of the genes under the influence of the local cytoplasm. Taken together, all these processes break up the originally fairly homogeneous system into a mosaic of physically and chemically different areas. We have called this the chemo-differentiation of the egg.

As a rule, chemodifferentiation does not manifest itself very clearly in the egg's external appearance. In exceptional cases, some of the determining substances may be visible externally because they have pigments of different colours; if so, their movements can be followed easily. In general, however, only experiments can inform us of the changes that have taken place in the egg.

Another process, however, also subsequent to fertilisation, and usually more or less simultaneous with the primary chemo-differentiation, can be easily observed. This is the *cleavage* of the egg, whereby the originally continuous mass of egg cytoplasm is divided into a number of cells. The zygote nucleus also divides a great number of times so that, in the end, the egg consists of a number of cleavage cells, or *blastomeres*, each containing one cleavage nucleus. A close correlation nearly always exists between the direction of the first cleavages, and the polarity and symmetry of the egg, as determined by its co-ordinate system. As a rule, the first two cleavages are in a meridional plane, and at right angles to each other, whereas the third cleavage lies in the equatorial plane.[1] From this point onwards, cleavage may follow several different courses. Sometimes it remains very regular so that the cleavage planes always cut through the egg in a definite direction, and two eggs of the same species in the same stage have identical structures. Evidently, the direction of cleavage is rigidly determined in such cases. The eggs of most molluscs and annelid worms may serve as examples (see above, p. 57). In other cases, cleavage soon becomes more or less irregular. The direction of the cleavage planes does not follow any general rule, and is apparently more or less independent of the polarity and symmetry of the egg. These differences are not essential for the further course of affairs, and they can be neglected here.

During cleavage the egg cytoplasm is distributed between the blastomeres. As a consequence of the previous chemo-differentiation, the cytoplasm of the various blastomeres will be physically and chemically different. We have already seen how this influences the realisation of the nuclear factors by the activation of different genes in different cells. We have also discussed (p. 69) how differences in the developmental potencies of the cells arise thereby. In other words, the germ now becomes a complex of cell groups with divergent potencies. The necessary conditions are thereby provided for the processes which now follow, in the course of which the embryo is formed (*embryogenesis*). This consists of two phases: the development of form, or *topogenesis,* and the differentiation of the tissues, or *histogenesis.*

Topogenesis is a series of movements, resulting in cell migrations, which bring the prospective tissues of the body

---

[1] **Ascaris** is an exception; here, the first cleavage is equatorial (Fig. 25). See also p. 76 on cleavage in insects.

to their appropriate places in the structural plan. These movements may take many different forms, such as folding, invagination, or shifting, either of cell groups, or of isolated cells.

In the eggs of most animals, a period sets in, directly after the completion of cleavage, in which these movements dominate the whole picture of development. So far, the egg has been rather inert, cleavage being its only manifestation of activity. Now, however, it suddenly seems to wake up, and to begin its real task, the formation of an embryo. The whole set of topogenetic phenomena immediately following cleavage is called *gastrulation*. Cleavage had already produced a vesicular germ, the *blastula*. During gastrulation this enters upon a phase in which some of the main components of the future embryo can already be recognized. We shall now discuss the process of gastrulation in two animal groups which have often been used for experiments in developmental physiology, viz. in sea urchins and amphibians.

The sea urchin germ at the end of cleavage is a vesicle with a wall consisting of a single layer of cells (Fig. 31a). The cells of the vegetative side are slightly higher than those of the animal pole. The latter carry a long tuft of stiff cilia, whereas the other cells have short, motile cilia. The beginning of gastrulation is marked by an indentation at the vegetative pole of the blastula. In this region, a certain number of cells disengage themselves from the epithelium, and move into the cavity of the blastula. They constitute the so-called *primary mesenchyme* (Fig. 31b, c). The indentation at the vegetative pole is shallow at first, but it penetrates gradually into the interior, thereby forming the *archenteron* (Fig. 31d, e). The opening which connects the archenteron with the exterior is the *blastopore*. The cells which form the wall of the archenteron are called the *endoderm*; to the cells remaining at the surface of the germ the term *ectoderm* is applied. When the archenteron has attained a length of about ⅓ to ½ the diameter of the blastula, certain cells at the tip of the archenteron begin to send out long thin pseudopodia, which attach themselves to the inner side of the ectoderm in the animal region. By contraction of these pseudopodia the tip of the archenteron is pulled toward the animal pole. At the same time, the pseudopodia bearing cells pull themselves out of the epithelium of the archenteron wall, and come to lie in the cavity, forming the *secondary mesenchyme* (Fig. 31e; Plate IX).

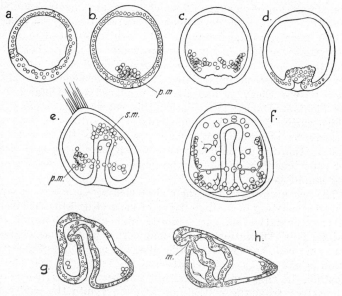

Fig. 31. The normal development of **Psammechinus microtuberculatus.** (*a*) early blastula; (*b-c*) formation of the primary mesenchyme (*p.m.*); (*d*) beginning invagination of the archenteron; (*e*) gastrula, from the left; formation of the secondary mesenchyme (*s.m.*); (*f*) older gastrula, dorsal view; formation of the skeleton; (*g*) the archenteron bends towards the ventral side; (*h*) formation of the mouth (*m*). After Schmidt.

Originally, the archenteron is straight, but later its terminal part bends, and comes into contact with the ectoderm (Fig. 31 g, h). At this point, an invagination, called the *stomodaeum*, is formed by the ectoderm. This breaks through into the archenteron, thereby forming the definitive *mouth*. The archenteron becomes the *gut*, the blastopore becomes the *anus.* The symmetry of the egg which, though preformed, was not yet visible, manifests itself in these processes. The embryo is now bilaterally symmetrical. The mouth has formed at its *ventral* side; the opposite side is *dorsal.* The cells of the primary mesenchyme accumulate in two groups, to the left and right of the gut. We have described above (p. 88) how the skeleton develops within these two cell masses (Fig. 31f). The calcareous rods of the skeleton show a rapid increase in length, and where their ends come into contact with the ectoderm, the latter bulges out, forming the so-called arms. In this way the blastula develops, via the *gastrula* stage, into a *pluteus* larva which already possesses the primordia of several organ systems characteristic of the sea urchins (Figs. 10, 11, 15, and 28). This development is achieved by a series of folding, invaginating and shifting processes.

In amphibians, gastrulation is somewhat more complicated. Cleavage here results in a blastula with a multi-layered wall. Its cavity (*blastocoel*) is eccentric, lying nearer to the animal pole of the egg. At this end, it is bordered by several layers of small cells, whereas the floor of the cavity is formed by several layers of large cells which are rich in yolk (Fig. 33a).

The size of the cells increases gradually from the animal pole towards the vegetative pole. The same statement applies to their yolk content. In many amphibians, the bilateral structure of the egg can be discerned in the blastula, because the grey crescent, which was formed soon after fertilisation, is still present. This marks the dorsal side of the egg.

Here, again, a complicated system of cell movements begins after the completion of cleavage, which shifts the cell material of the prospective organs towards their appropriate places. The best way to elucidate this complicated process is to consider the surface of the blastula to be divided into three parts by two imaginary lines. These parts are: (1) a *marginal zone* which surrounds the egg in the vegetative half, and is more or less parallel to the equator; it is wider at the dorsal side, and here it extends into the animal half; (2) the *animal field*, consisting of small, usually darkly pigmented cells; (3) the *vegetative field*, consisting of large, unpigmented cells, containing much yolk (Fig. 32). The boundaries of these areas cannot be distinguished in the egg by their outward appearance; nevertheless, their importance will appear in the later development.

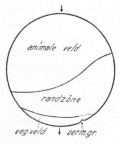

Fig. 32. Diagram of an amphibian blastula. *Animale veld* = animal field; *oerm.gr.* = blastopore furrow; *randzone* = marginal zone; *veg. veld* = vegetative field.

The beginning of gastrulation is marked by a small indentation at the dorsal side of the blastula, in the vegetative field, just under the grey crescent which here lies in the marginal zone. This indentation is called the *blastopore furrow*. It is produced by an active change in shape of the cells of that area, which become club-shaped and thereby move inwards. The depth of the indentation soon increases, and in this way the *archenteron* extends into the egg, along its dorsal side, and in the direction of the animal pole (Fig. 33 a-e). The invagination then begins to spread laterally, more or less along the boundary between marginal zone and vegetative field. All along this boundary, marginal zone material begins to roll in, forming a protruding edge, called the *lip* of the blastopore (Fig. 36 a, b). The invaginating archenteron gradually compresses the blastocoel. In the course of this invagination, material which was originally on the surface is carried into the interior, so that the wall of the archenteron consists of cells which initially lay at the surface. Material of the vegetative field as well as material of the neighbouring dorsal part of the marginal zone is rolled in over the lips of the blastopore. The vegetative material will form the ventral wall of the archenteron, its so-called *floor*, whereas the material from the marginal zone will form its *roof*. In the end, gastrulation leads to an almost complete disappearance of the material of the vegetative field into the interior. Only a small part of this material still protrudes from the blastopore as the *yolk-plug* (Fig. 33 c, d). Finally this, too, disappears, and the blastopore lips, approaching each other

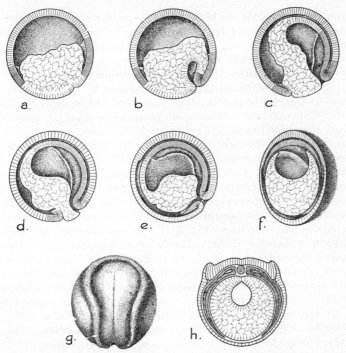

Fig. 33. Diagram of gastrulation in amphibians. (a) blastula, halved in the median plane, with thin roof and thick floor (coarse hatching: animal field; fine hatching: marginal zone; the outline of cells is indicated in the vegetative material), blastopore furrow below at right; (a-e) invagination of the archenteron, starting from the blastopore furrow; inward movements of the vegetative material; rolling in of the marginal zone material over the edge of the blastopore; (f) cross section through the same stage as in (e), the trough shaped endoderm forms the floor and the sides of the archenteron, the prospective notochord and mesoderm its roof, whereas the ectoderm covers the outer surface of the germ; (g) dorsal view of neurula, showing the neural plate, surrounded by a raised fold which separates it from the rest of the ectoderm; (h) cross section through the same stage as in (g); the endoderm has closed dorsally, forming the gut, the archenteron roof has differentiated into notochord and mesoderm, the ectoderm into neural plate and skin ectoderm. After Hamburger and Mayer.

from all sides, close the blastopore. The material of the marginal zone also disappears from the surface by rolling-in over the lips of the blastopore. At the same time, the material of the animal field expands strongly, and finally covers the whole surface of the germ. It also becomes thinner, and the multi-layered arrangement of the cells is replaced by a single-layered one. At the end of gastrulation, therefore, the embryo consists of (1) an external *ectoderm*, or outer germ-layer, originating from the animal field, (2) an internal *endoderm*, or inner germ-layer, forming the floor and the sides of the archenteron, and arising from the original vegetative field, and (3) the roof of the archenteron, formed by the invaginated marginal zone material, and constituting the middle germ-layer, or *chorda-mesoderm*. The archenteron roof, consisting of the prospective notochord and mesoderm sheets, extends ventrally between ectoderm and endoderm (Fig. 33f). The main axes of the embryo have now become visible: The blind end of the archenteron marks the anterior end of the embryo; here the mouth will break through. The blastopore lies at the posterior end, and will become the anus. The back of the embryo will be formed at the dorsal side of the germ, and its belly at the opposite side.

At the conclusion of gastrulation, topogenesis has not yet come to an end. The next phase, known as *neurulation*, follows immediately. During this period, new movements result in the formation of the first organ systems of the embryo from the material brought into place by gastrulation. The dorsal ecto-derm thickens and becomes the *neural plate* (Fig. 33g). The edges of this plate fold upwards (*neural folds*), and finally fuse, forming the *neural tube*. This tube is the primordium of the embryonic central nervous system. The rest of the ectoderm is the prospective *epidermis*. The mid-dorsal cells of the archenteron roof, under the middle of the neural plate, become free from the rest of the roof, and form a longitudinal cylin-drical rod, the *notochord*. The adjoining strips, to the right and left of the notochord, first thicken, and then divide into a number of segments, the so-called *somites,* lying in a longi-tudinal row. These will give rise to the musculature of the trunk, and to mesenchyme, from which the axial skeleton develops.

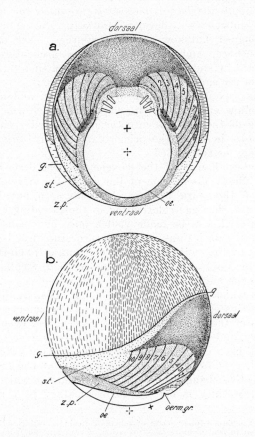

Fig. 34. Diagrammatic projection of the organ primordia of a urodele embryo on to the blastula surface, seen (*a*) from the vegetative pole, and *(b)* from the left side. Coarse hatching = epidermis, fine hatching = neural plate, closely dotted = notochord, sparsely dotted = mesoderm, white = endoderm; future somites numbered; *g*: boundary of the invaginated area; *oe*: future blastopore lip; *oerm.gr.*: blastopore furrow; *st.*: tail area, *z.p.*: lateral plate mesoderm. After Vogt.

The remaining, lateral and ventral parts of the mesodermal sheets supply the *lateral mesoderm* in which the body cavity of the embryo will develop. Initially, the endoderm was trough-shaped, but now its edges approach each other under the notochord, and fuse so that the archenteron is now surrounded by endoderm on all sides. From then on it becomes the *gut* of the embryo (Fig. 33h).

In this way, the processes of gastrulation and neurulation, which follow each other without delay, transform the simple structure of the blastula into the complicated architecture of the embryo, in which the primordia of several organ systems can already be recognised.

Only a small part of the movements here described can be seen by direct observation of the germ. The development and closing of the neural plate can be easily seen, but the shifting processes during gastrulation, as in the infolding of marginal zone material, can be made visible only by means of special techniques. Our insight into these processes has been greatly deepened by the method of colour marking *intra vitam*, invented by Walter Vogt (1925-29). In this method, small marks are made on the surface of the germ with so-called "vital" stains, i.e. dyes which do no damage to living cells. In this way, we can investigate the particular organs of the embryo into which the cells of the various areas of the germ will eventually find their way, and these organs of the embryo can, so to speak, be projected back on to the surface of the blastula (Fig. 34).

All movements involved in gastrulation and neurulation are due to active changes in shape of the cells of which the germ consists. Motility is one of the elementary properties of all

PLATE IX. The cellular basis of morphogenesis. Cellular activities during gastrulation in sea urchins. (a) Formation of pseudopodia from tip of partly invaginated archenteron. (b) The pseudopodia attach to inner wall of blasto-coel. (c) By contraction of the pseudopodia the archenteron is drawn inwards. Row of primary mesenchyme cells at right. (d) Tip of archenteron with pseudopodia. At upper left cone of attachment in the ectoderm pulled inward by tension exerted by contracting pseudopod. (e) Same embryo as c, about 10 minutes later. Note progress in invagination, and large ventral exploring pseudopod at right. (f) Bridge of secondary mesenchyme cells pulled out by pseudopods from the archenteron tip towards the ventral side (region of future stomodaeum). Note cone of attachment. (Partly from H. Kinnander and T. Gustafson, **Exp. Cell Res.** 19, 1960; T. Gustafson and H. Kinnander, **Exp. Cell Res.** 21, 1960) (Courtesy of Prof. T. Gustafson).

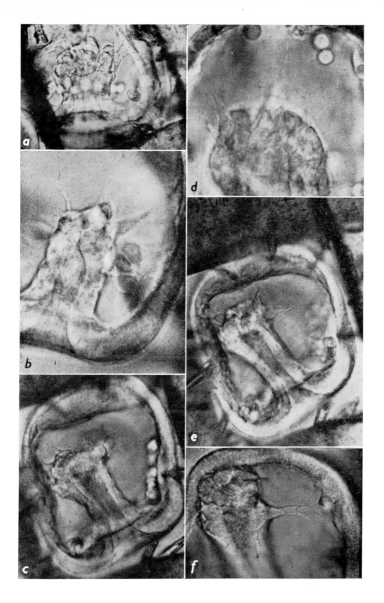

PLATE IX.

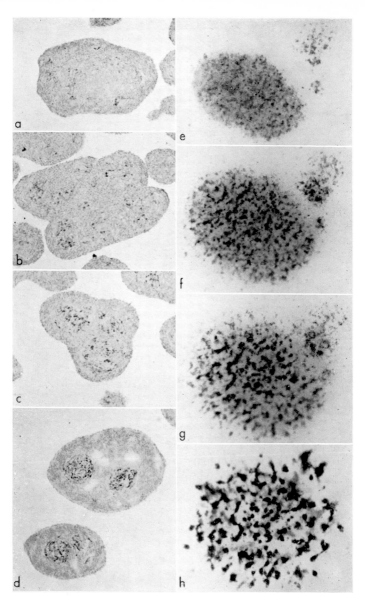

PLATE X.

cells. Most cells are capable of active changes in shape, although not all cells show this so clearly as the elements of the musculature. Many animal eggs, too, are motile to a lesser or greater extent. In some lower animals, such as the sponges, the eggs can crawl around freely in the maternal tissues. These eggs are very similar to amoebae. In other animals, the eggs show amoeboid movements only at certain times, e.g. in the snail *Limnaea* immediately after the expulsion of the polar bodies. We can imagine that this power of locomotion becomes specialised in divergent directions in the various cell groups of the germ, due to the modifications in the physico-chemical condition of the parts of the egg, caused by chemodifferentiation. The cells of one area will show a tendency to expand parallel to the germ surface, whereas in another area they tend to extend at right angles to it. In yet another area they may assume a club-shape by enlargement of the part pointing towards the centre of the germ, and narrowing of the other end. This will cause the cell to move into the interior, and thereby give rise to an invagination of the superficial material. It has indeed been proved experimentally that different parts of the early gastrula of amphibians show tendencies for different movements. Part of the marginal zone tissue of one gastrula can be grafted into the ventral side, or into the animal field, of another early gastrula. A blastopore will then be formed by invagination in the graft, all of which is rolled into the interior, often together with a number of cells of the host. This

---

PLATE X.   Type-specific sorting-out of cells in mixed cell aggregates. (a-d) Segregation of chick embryonic heart cells from chick embryonic retinal cells; sections of aggregates of various ages. Heart cells contain darkly stained, clumped glycogen granules. At 17 hours (a) the heart cells have left the free surface and are starting to form minute clusters. By 24 hours (b) fusion of heart clusters has produced internal islands of heart tissue. Fusion of heart islands is seen at 31 hours (c). Eventual distribution of tissues is approached by 66 hours (d). (e-h) Direct observation of cellular segregation in living mixed cell aggregates of chick retinal pigment cells and heart ventricle cells. Pigment cells black. (e) 6 hours in culture. Sorting out has already resulted in the formation of distinct small clusters of pigment cells. (f) 20 hours in culture. The aggregates have spread over the substratum. Pigment cell clusters have enlarged and fused with adjacent clusters. (g) 34 hours in culture. Sorting out and fusion of clusters has continued. (h) 5 days in culture. Changes in the form of pigment cell clusters have continued. Clusters have become more compact. a-d after M. S. Steinberg, in **Cellular Membranes in Development,** Acad. Press 1964; e-h after J. P. Trinkaus and J. P. Lentz, **Dev. Biol.** 9, 1964 (Courtesy of Prof. M. S. Steinberg and Prof. J P. Trinkaus).

Fig. 35. "Hyperblastula" of **Triton**, formed by the animal half of an egg. Strong expansion of the animal material leads to folding. After Ruud.

shows that marginal zone material must have the potency to roll in actively, simultaneously extending to form an archenteron. On the other hand, the material of the animal field shows a tendency to expand parallel to the surface. Spemann (1931) cut two blastulae in halves along the equator, and grafted the two animal halves together. This resulted in a germ with an excessively expanded ectoderm, forming many loose folds and flaps (Fig. 35). This can be explained as follows. In normal development, the expanding ecto-derm takes the place of the marginal zone material that is rolled in during gastrulation. But, in the germ resulting from the fusion of two animal halves, no rolling in takes place, and there is therefore no room for the expanding ectoderm, which is obliged to fold. The material of the vegetative field shows hardly any capacity for active movement. It seems to be more or less passive in gastrulation; only the very first beginnings of the blastopore furrow in this area appear to be due to an active power of invagination of the cells, which assume a club shape, as mentioned above (Vogt, 1922, 1929).

This power of active movement, or *topogenetic potency* of the cells, is very clearly shown if the normal co-operation of the migratory tendencies in gastrulation is disturbed in some way or other. Holtfreter (1933) obtained such a disturbance in newt eggs by taking them out of their capsules at an early stage, and transferring them into certain salt solutions. The resulting disturbance of gastrulation is known as "exogastrulation". No invagination of the archenteron, accompanied by rolling in of the material of marginal zone and vegetative field takes place. On the contrary, an evagination occurs at the edge of the blastopore, because marginal zone material moves through the blastopore furrow to overlie the vegetative field. The result is a germ in which endoderm and chordamesoderm are not surrounded by, but lie altogether outside the ectoderm. The latter

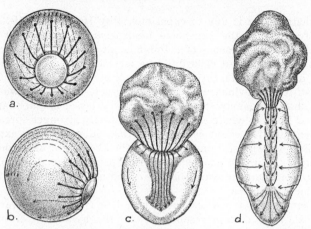

Fig. 36. (*a-b*) normal gastrulation, seen *a*: from the vegetative side, and *b*: from the left side; (*c-d*) exogastrulation. The arrows indicate the directions in which the material moves. After Vogt and Holtfreter.

forms an empty, folded pouch which may eventually become completely separated from the rest of the germ at the blasto-pore edge (Fig. 36). Now it is a remarkable fact that all the other movements of gastrulation, and also part of the sub-sequent movements of neurulation, take place in the normal way, even though invagination has been suppressed, and turned into an evagination. Notochord and somites develop from the chordamesoderm, although the latter is not surrounded by ectoderm. There is a beginning of differentiation into head, trunk, and tail, and even the first primordia of branchial pouches are formed in the endoderm of the head-region. This shows that the various topogenetic processes are more or less mutually independent. It is essential for normal gastrulation, however, that these processes are linked together in an orderly fashion. Schechtman (1942) has demonstrated that normal rolling in of prospective notochord material, and closure of the blastopore, will only occur if the connections between the parts of the marginal zone are uninterrupted.

Much of our insight into the nature of the moving forces

in gastrulation is due to experiments by Holtfreter (1943-44). He found that the expansive tendencies of the animal field depend on the presence of a tough, semi-elastic surface "coat". This coat unites all cells into one mechanical unit, and in this way assists in integrating the collective cell movements. In so far as it is provided with a similar layer, the vegetative field also shows expansive tendencies, but to a lesser extent than the animal field. In whole embryos the relatively stronger spreading tendency of the ectoderm prevents the endoderm from expanding into a peripheral epithelium; in the absence of an epidermal covering, as in the exogastrulae, the spreading tendency of the coated endoderm prevails, however, and no internal archenteron can be established.

A similar coat has also been found in fishes. Moreover, the coat shows a certain resemblance to the hyaline layer in sea urchins, which has a similar function in binding the cells together. Both the coat and the hyaline layer are dissolved in Ca-free media. In sea urchins the hyaline layer is not indispensable for the invagination of the archenteron, however (Moore, 1952).

Certain specific attractions and repulsions existing between different types of cells also result in topogenetic processes. Holtfreter (1939), who studied these phenomena, has summarized them under the term "tissue affinity". These forces manifested themselves clearly in experiments in which different tissue elements were cultivated together. Holtfreter found that under their influence parts of a cell mass may become separated, or, conversely, cell groups of different types may fuse and interpenetrate, so that they come into close contact. Such processes play a very important role, particularly in later development when the organs of the embryo begin to differentiate (Townes and Holtfreter, 1955).

This subject has recently been further studied in experiments, in which the reaggregation of isolated cells, obtained by disaggregating tissues with special methods breaking down the primary adhesions between cells, was observed. When suspensions of disaggregated cells of different types are mixed, the cells first reaggregate into clumps which gradually fuse into larger masses. The initial clumps consist of cells of different type, but after their fusion to larger cell masses the cells within a mass begin to sort themselves out into regions made up of a single type of cells, often arranged in a regular order. Ultimately this may sometimes lead to the differentiation of complex organs of fairly normal structure (Plate X).

The main morphogenetic events in early sea urchin development: blastula formation, invagination of the archenteron, immigration and further distribution of the primary and secondary mesenchyme cells, development of the skeleton, formation of the mouth, etc., can be accounted for in terms of a few basic cellular activities, such as cell adhesion and pseudopodial activity (Gustafson and Wolpert, 1963) (cf. Plate IX). The time-space pattern of these cellular activities appears to be relatively simple, and related to the primary co-ordinate system of the egg.

The following investigations have thrown some light on the problem of the causation of the different topogenetic potencies in different parts of the blastula. It has been known for a long time that the gastrulation of amphibian eggs is often disturbed if, soon after fertilisation, the eggs are fixed with the animal pole downwards (Schultze, 1894) (cf. p. 52). Schleip and Penners (1925-28), and Motomura (1935) repeated this experiment, and finally Pasteels gave a satisfactory analysis of the phenomenon (1938-39). It appeared that in such eggs a blastopore may be formed by invagination at any point where yolk material is in direct contact with the egg surface, but preferably at the side directed toward the original centre of the grey crescent. From these observations, Dalcq and Pasteels concluded that the occurrence of the potencies for gastrulation in normal eggs is governed by (1) the yolk gradient, and (2) the cortical field (see above, p. 52). At any point, both the *"yolk factor"* V, and the *"cortical factor"* C, have a definite value. Dalcq and Pasteels thought it probable that the product of these factors, $C \times V$, which also has a definite value at each point of the egg surface, was a measure of the differences in topogenetic potency. Where the product $C \times V$ exceeds a certain value, the cells will acquire a tendency to roll in. Where it remains below the liminal value, there is a tendency for expansion in all directions. Where the yolk concentration rises above a certain absolute value, the cells remain more or less inert. Dalcq and Pasteels succeeded in giving an elegant derivation, based on their hypothesis, of the distribution of the topogenetic potencies in the normal blastula (Fig. 37).

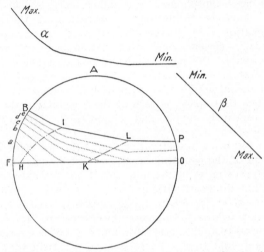

Fig. 37. Diagram of the interaction of the gradients in the development of amphibians. The cortical factor C decreases from dorsal side (left in the figure) to ventral side (on the right), according to line α. The yolk factor V decreases from the vegetative pole (below) to the animal pole (above), according to line β. The product $C \times V$ is maximal at F (centre of the dorsal marginal zone), and decreases gradually from there. The lines a, b, c, d, and e indicate planes in which the product has the same value. The lines HI and KL are the boundaries of areas in which the ratio C/V has the same value. After Dalcq and Pasteels.

We can summarise the above as follows. The chemodifferentiation which occurs under the influence of the egg cortex gives rise to different topogenetic potencies in the various parts of the egg. At a given moment, these lead to the occurrence of cell migrations which result in gastrulation, and further topogenesis.

CHAPTER VIII

Induction and organisation

I. Neurulation

The cell migrations during gastrulation, and the further topogenetic processes to which the cells are subjected, bring the
materials for the embryo's future organs into those places
where they belong according to its structural plan. But they
also have another consequence. Cell groups which originally
were separated by considerable distances, and which, as a
result of chemodifferentiation and differential gene activation,
differ in physical and chemical condition, now become immediate
neighbours. It may be expected, therefore, that they will now
influence one another, and that the new topographical relations
between the cell groups will in this way initiate new processes
of chemodifferentiation. The outcome of these topogenetic
processes will thus be a considerable increase in the spatial
multiplicity of the embryo.

There are, indeed, some observations, which give clear
evidence of changes in the physico-chemical constitution of the
cells during gastrulation. Woerdeman (1933) was the first to
show that a sudden fall occurs in the glycogen content of the
glycogen-rich marginal zone cells of amphibians, at the moment
when these cells are rolled in over the lip of the blastopore.
The following experiments proved that there was a causal
connection between the processes of invagination and of
glycogen decomposition. First of all, dorsal marginal zone
material of an early gastrula was grafted into the ventral marginal zone of another germ. It then became rolled in, either over the
ventral lip of the host blastopore, or over the edge of a secondary blastopore (see p. 107). The moment of disappearance of
the glycogen always coincided with that of rolling in. Secondly,
cells of the animal field were grafted into the dorsal marginal

113

zone. In this material, too, glycogen decomposition took place when it was rolled in over the lip of the blastopore, even though in normal development it would have stayed at the surface, and retained its glycogen. Thirdly, the same material did not lose its glycogen if brought directly into the interior of the germ through a slit cut in the roof of the blastula (Raven, 1933-35). Finally, marginal zone material does not lose its glycogen when cultured in isolation from the rest of the embryo (Jaeger, 1945). Hence, the disappearance of the glycogen presumably is a direct consequence of the rolling in of the cells during gastrulation.

In this context mention must also be made of observations on the distribution of ribonucleic acid and sulphydryl compounds (Brachet, 1940-42), and of alkaline phosphatase (Krugelis, 1947) in amphibian embryos. In each case, a more homogeneous distribution, obtaining in the younger stages, was seen to be replaced during gastrulation by a distribution in which there were distinct differences in concentration among the various parts of the germ.

However, such direct observations on changes in the physicochemical constitution of the cells are still rather fragmentary. In most cases, the occurrence of chemodifferentiation cannot be demonstrated directly; it appears only indirectly from changes in the potencies of the cells during gastrulation. We possess more accurate knowledge on this point, particularly in amphibians.

The *histogenetic potencies,* or potentiality for differentiation, of the cells can be studied by removing a small group of cells from the germ, and rearing it in isolation in an appropriate salt solution. This method is called *explantation. Transplantation,* the grafting of the removed cell group into another embryo, is another useful method. In the case of explantation, we may generally assume that the cells are not exposed to any specific modifying influences on the differentiation processes. Such experiments inform us what the cells concerned can do "on their own", and therefore give us an insight into their *autonomous powers of differentiation.* Cells transplanted into another embryo will there be subjected to the action of the surrounding tissues of the host. This will vary according to the place and time of grafting, so that divergences in different-iation will occur. By grafting similar cell groups into various parts of hosts of different nature and age, the *range of possible differentiations* of the material can be studied.

The differentiation potencies of various parts of young amphibian gastrulae have been studied by these methods. The following results were thereby obtained. The material of the vegetative field has a very strong capacity for self-differentia-tion already. Holtfreter (1931) cultured this material in a salt solution, and found that it differentiated into vesicles with an epithelial wall. The structure of this epithelium varied in accord-ance with the place of origin of the explant. Sometimes it was similar to the epithelium of the branchial part of the gut, at other times it had the character of stomach or mid-gut epi-thelium. This shows that there are qualitative differences in differentiation potency even within the vegetative field. On the other hand, the range of possible differentiations of this material is limited. If transplanted it produces only endo-dermal tissues. This proves that the cells' inherent *"differentia-tion tendency"* can maintain itself, in spite of environmental influences that work in another direction.

The marginal zone material also appears to have an intrinsic capacity for differentiation. In Holtfreter's explantation ex-periments, it differentiated into notochord, muscle, kidney epithelium, etc. Within the marginal zone, too, there are qualitative differences in differentiation potencies. The tendency to form a notochord preponderates in the dorsal part, that to form muscular tissue in other places, etc., as proved for example by Bautzmann's (1933) transplantation experiments. But the mode of differentiation of the various parts of the marginal zone is not yet irrevocably fixed, or "determined". In Holtfreter's experiments, notochord and muscle nearly al-ways arose simultaneously from the same material. If the explant is large, it may even produce ectodermal tissues, as was first noted by Lopashov (1935). Evidently, the marginal zone material has a wider range of possible differentiation than is manifest in normal development.

Hence, the cells of the vegetative field and the marginal zone, at the beginning of gastrulation, possess inherent capacities for self-differentiation, which even show local qualitative differences. The animal field, on the other hand, gives an entirely different picture. When isolated in a salt solution, material from this area produces only irregular masses of cells, in which the individual cells have an atypical structure. Only if mesenchyme cells are present as well, vesicles will be formed, the walls of which consist of an epithelium that is completely similar to the epidermis of normal embryos. This shows that here the capacity for self-differentiation is still very poor. On the other hand, the range of possible differentiations is still practically unlimited. A wide variety of transplantation experiments has shown that the cells of the animal field can differentiate either into muscle, or into nervous tissue, into notochord, or kidney, gut or liver. In other words, practically all the tissues of the body can be formed by this material. It is still "omnipotent" (Holtfreter, 1933; Raven, 1935). There are no local differences within the animal field in this respect. Cells from the dorsal half, i.e. material of the prospective neural plate, behave in entirely or practically the same way as cells from the ventral half, i.e. prospective epidermis (cf. Fig. 34). If parts from these two regions of the animal field are exchanged, they fit harmoniously into their new environments, and can entirely substitute for each other (Spemann, 1918), (Fig. 38). Evidently the cells of the animal field are of a highly indifferent character; their chemodifferentiation is still very slight.

Now it is in this material that the progress of chemo-differentiation during gastrulation, and the modification of differentiation potencies involved, are manifested most clearly. For a completely different situation is found if after gastrulation, at the neurula stage, the potencies of the ectoderm are again investigated by means of transplantations and explantations. At this stage the ectoderm cells are no longer alike in their differentiation potencies. A part of the neural plate will produce nervous tissue, even if grafted into skin ectoderm (Fig. 47), and, conversely, a piece of skin ectoderm, grafted into the foreign environment of the neural plate, will still produce epidermis (Spemann, 1918). Also in the case of explantation, neural plate material will differentiate into nervous tissue, and skin ectoderm (in the presence of mesenchyme cells) into epidermis (Holtfreter, 1931). Evidently, the two regions of the ectoderm have now acquired different differentiation potencies. At the same time, the range of possible differentiations has been cut down. After transplantation into a foreign environment, ectodermal grafts will produce only ectodermal

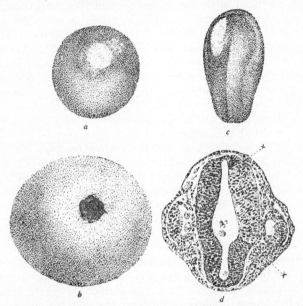

Fig. 38. Heteroplastic transplantation of presumptive ectoderm at an early gastrula stage. (a) gastrula of **Triton taeniatus** with graft from **T. cristatus**; (b) gastrula of **T. cristatus** with **taeniatus** graft; (c) the **taeniatus** embryo of fig. (a), at neurula stage; the graft (light pigmentation) occupies the front part of the neural plate; (d) cross section of this **taeniatus** embryo at a later stage; part of the brain wall (between the crosses) has developed from the light **cristatus** graft, which has produced brain tissue in accordance with its new environment. After Spemann.

tissues, and no muscle tissue, notochord, or gut any more (Mangold, 1923). This shows that chemodifferentiation has progressed in the animal material, the parts of which were still equivalent at the beginning of gastrulation. The cells have been fixed in one definite direction; they have become *determined* to one definite differentiation.

How does this determination take place? Experiments by Spemann (1918) have given a clear answer to this question. It appears that the material of the dorsal part of the marginal zone is provided with special properties. We have seen that during gastrulation this material is rolled in over the dorsal lip of the blastopore, and subsequently it forms the roof of the archenteron, from which the notochord and somites originate. Spemann removed part of this material from an early gastrula, and grafted it into the ventral side of another germ of the same age. In this region the primordium of a complete embryo was formed, consisting of neural tube, notochord, somites, and gut. Spemann's pupil Hilde Mangold (1924) repeated this experiment, but combined two different species, in a so-called *heteroplastic transplantation*. The graft was taken from the dorsal lip of the blastopore of *Triton cristatus*, and it was grafted into the ventral side of *T. taeniatus*. It became clear in this case that only part of the secondary embryo originated from the graft itself. The latter was, indeed, rolled in under the influence of its special topogenetic potency, and it produced notochord and somites. The neural primordium of the secondary embryo, however, did not arise from the graft, but entirely or for the greater part from the overlying ectoderm of the host. The secondary gut primordium also developed from host material (Fig. 39). This means that the graft enforced a definite course of development on the cells of the host. It *"induced"* them to undergo a certain differentiation. This, however, did not lead to a chaotic aggregate of tissues, but the various organs produced by the co-operation of graft and host showed normal topographical interrelations so that together they constituted an organised embryo. In very favourable cases, this may go so far that the anterior end of the secondary embryo forms a distinct head with brain, eyes, ear-

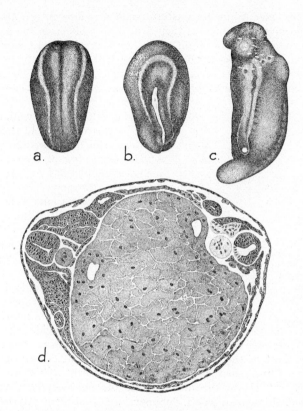

Fig. 39. Organiser activity in amphibians. (*a*) neurula of **Triton taeniatus**, with (*b*) a secondary neural plate at the ventral side, induced by a piece of dorsal marginal zone material from **T. cristatus**, part of which is still visible as a narrow white band in the neural plate; (*c*) embryo of **T. taeniatus** with a secondary embryo primordium (with neural tube, somites, ear-vesicles, and protruding tail-bud) in its left side; (*d*) cross section of the embryo of fig. (*c*), showing primary embryo primordium on the left, secondary primordium on the right; the latter consists partly of the light tissues of the organiser, partly of the dark material of the host. After Spemann and H. Mangold.

vesicles, etc., while its posterior end develops into a tail bud. The secondary embryo is then distinguishable from the primary embryo only by its slightly smaller size (Pl. XI).[1] For these reasons, Spemann called such a graft an *"organiser"*, and the dorsal marginal zone from which it originated the *"organisation centre"* of the embryo. In this way he voiced his conviction that this area played a role in normal development as well. Bautzmann (1926) determined the exact limits of the organisation centre. At the beginning of gastrulation, the organising power, or, more accurately, the power to induce a neural plate, was found to be limited to the dorsal half of the marginal zone, i.e. the area which in normal development provides the notochord and somites (Fig. 40).

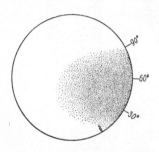

Fig. 40. Extent of the organisation centre (dotted) in the initial stage of gastrulation of **Triton taeniatus.** Germ seen from the left; the wavy line indicates the blastopore furrow. After Bautzmann.

The first experiments of Spemann and Hilde Mangold had already shown that in secondary embryos the organ boundaries did not necessarily coincide with those between graft and host. In some cases, part of the graft was not rolled in, but remained at the surface, and associated with the ectoderm. This was then incorporated into the neural plate of the secondary embryo, so that this organ was a "chimera", consisting partly of graft tissue, and partly of host tissue (Fig. 39 b, d). A similar chimerical composition was sometimes encountered in the notochord and somites of the secondary embryo, where cells of the host's ventral mesoderm, bordering on the graft, took part in the formation of these organs (Fig. 39 d). This means that the inductive activity of the graft did not only manifest itself in the overlying ectoderm, and the underlying endoderm; it had evidently also extended from cell to cell within the meso-

[1] Facing page 122.

derm, and forced a development in an aberrant direction on the mesoderm of the host. It is important to make a clear distinction between these two types of embryonic induction, and to discuss them separately. They are: (1) *induction by contact*; here the *"inductor"* and the material that reacts to it, the so-called *"reaction system"*, are in immediate contact, and

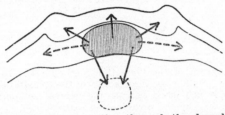

Fig. 41. Diagram of a cross section through the dorsal part of an embryo, illustrating the difference between contact induction and progressive induction. The organiser (hatched) influences the overlying ectoderm and the underlying endoderm by contact induction (arrows drawn in full), but it influences the surrounding mesoderm by progressive induction (interrupted arrows).

(2) *progressive induction*, which extends from cell to cell in a coherent layer of cells (Fig. 41).

In normal development, the material of the dorsal marginal zone, or organisation centre, becomes invaginated during gastrulation, and forms the archenteron roof. It becomes located under that precise part of the ectoderm which will later differentiate into the neural plate. The hypothesis is obvious that here, again, "contact induction" plays a role, and that the neural plate is determined under the influence of the underlying archenteron roof. Marx (1925) demonstrated that the latter material does indeed possess inductive powers. He cut out part of the archenteron roof of an older gastrula, and brought it into the blastocoel of a blastula. The slit in the blastula wall through which it was inserted soon healed again (Fig. 42 a). During the gastrulation of the host, the blastocoel was obliterated by the archenteron, and the implant then lay immediately under the ectoderm. Therefore, it was in a position to exert such inductive influence as it possessed (Fig. 42 b). This

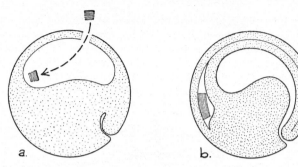

Fig. 42. Diagram of the "insertion method"; (a) through a small incision, the implant is inserted into the blastocoel; (b) during gastrulation it is pushed against the ventral ectoderm of the host by the invaginating archenteron.

"insertion method" has proved very effective, and it has been employed in a great number of other investigations on the inductive powers of certain implants. In Marx's experiments, small but distinct neural plates developed in the ectoderm overlying the grafts consisting of archenteron roof, proving that this material possessed inductive powers.

In this way the mechanism of the determination of the neural primordium in normal development was elucidated. Position and size of the neural plate evidently depend on position and size of the underlying archenteron roof. Moreover, Spemann (1931) demonstrated the existence of qualitative differences within the organisation centre, which are responsible for the regional differentiations of the neural plate. The first invaginated part of the marginal zone, which becomes located at the anterior end of the archenteron roof, induces especially the cranial parts of the neural primordium, i.e. the brain, with eyes and ear-vesicles. The material that invaginates later, and forms the caudal parts of the archenteron roof, induces the caudal regions of the neural primordium, i.e. the spinal cord (Pl. XII).[1] It appeared that the influence of the archenteron roof extended to the further development of the neural primordium as well. Topogenesis and differentiation of the central nervous system

[1] Facing page 123.

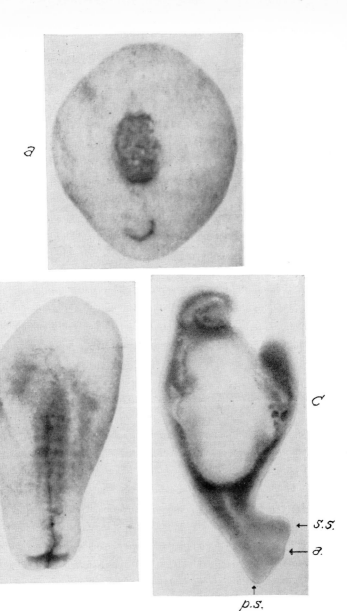

PLATE XI. The organiser and its activity. (a) Gastrula of **Triton** with implanted organiser (dark pigmentation) opposite the blastopore lip. (b) The same embryo at the neurula stage, (ventral view); a secondary neural plate has been induced. (c) The same embryo a few days later. A secondary embryo primordium (right), with head and tail-bud has been formed at the side opposite to the primary primordium. Both embryos have a common anus (a), situated between the primary (p.s.) and secondary (s.s.) tail-buds. After Spemann.

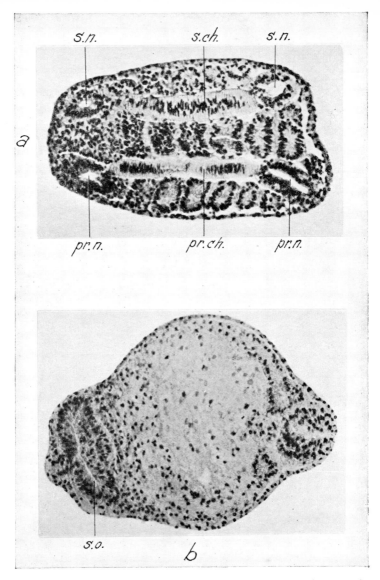

PLATE XII. Head- and trunk organiser. (a) Part of the blastopore lip of an older gastrula has induced a secondary primordium, of a trunk-like nature, in the trunk of the host. The two embryos are parallel. In both, the notochord is sectioned lengthwise (**pr.ch.** and **s.ch.**), whereas the neural tube is sectioned at two places (**pr.n.** and **s.n.**). The somites of the two embryos may be seen on both sides of each notochord. (b) Part of the archenteron roof of an older gastrula has induced a piece of fore-brain, with eye-vesicles (**s.o.**), in the trunk area of the host (cf. Fig. 46a). After Spemann.

proceed in immediate dependence on the underlying notochord and somites. Many experiments by Lehmann (1926-35), Holtfreter (1933), and others have demonstrated that defects in the archenteron roof cause characteristic abnormalities in the brain and spinal cord. Further, it was shown that the chordamesoderm retains its inductive powers for a long time. Pieces of notochord or somites from older embryos (neurula or tailbud stage) have been investigated by the "insertion method". They proved still to be able to induce a neural plate in the ectoderm of the gastrula (Bautzmann, 1928; Holtfreter, 1933). In contrast, the reactivity (*competence*) of the ectoderm to stimuli inducing a neural plate was found to be restricted to a certain stage only. Once the germ has reached the neurula stage, the reactivity has disappeared. This is the reason why the secondary neural plate nearly always becomes visible simultaneously with the primary neural plate, even though the inducing stimulus has had an opportunity to exert its influence long before that.

The inductive activity is not species-specific. In the original work of Spemann and H. Mangold an organiser of *Triton cristatus* was already found to be active after heteroplastic transplantation into *T. taeniatus*. Moreover, induction was found to take place when two more distant species were combined (so-called *xenoplastic* transplantation). In Geinitz' (1925) experiments, marginal zone material not only from axolotl and *Pleurodeles,* but even from frogs (*Rana fusca* and *R. esculenta*) and toads (*Bombinator*) induced neural plates in the ectoderm of *Triton*.

Even greater interest was aroused by the discovery that the power of inducing a neural plate was possessed, not only by the roof of the archenteron, but also by other tissues, which are not in normal development involved in the formation of the neural plate. This was demonstrated by bringing such abnormal inductors into immediate contact with overlying gastrula ectoderm by means of the "insertion method" mentioned above. Inductive powers were found not only in the neural plate itself, and in brain tissue of older larvae (Mangold and Spemann, 1927), but in material of the larval limb primordia as well!

(Mangold, 1928). Next, Holtfreter (1934) showed that many tissues of a wide variety of animals, both vertebrate and invertebrate, have the power of induction. He studied not only living tissues, but also dried, boiled, and minced material, and these, too, appeared to have inductive capacities in many cases. Moreover, not only neural plates, but also other organs, such as olfactory pits, eyes, ear-vesicles, limbs, etc., were formed by the host ectoderm under the influence of the inductors. It was concluded from these experiments that the inducing agent, or possibly agents, are very widespread in animal tissues, and that the processes concerned are of a very general character, and of low specificity.

All interest now centred on the problem of the nature of induction. Marx (1931) found that narcotized organisers did not lose their power of induction. Spemann (1931) found the same in ground organisers. Even dead tissue proved still to possess inductive powers. Organisers that had previously been dried, frozen, boiled, or killed with alcohol, still caused the formation of neural plates in the overlying ectoderm (Holtfreter, 1933). The remarkable discovery was made that even embryonic tissues which had no power of induction when alive, acquired this capacity after death. This was found, for example, in the prospective ectoderm and endoderm of gastrulae.

It was an obvious hypothesis at this stage to ascribe the induction to a substance, given off by the inductor, which causes the ectoderm cells to form neural tissue. This inducing substance was called *"evocator"* by Waddington, and *"organisine"* by Dalcq. In normal development it would be given off only by the archenteron roof. It might be present in other parts of the embryo, but there it would be prevented from taking effect in some way or other. It was assumed that many other animal tissues also contained the substance. This was confirmed by the successful preparation of extracts from animal tissues which, absorbed by an agar-agar carrier, induced a neural plate in the ectoderm of an amphibian gastrula. In principle, the determination of the nature of these evocators had thereby become possible. On this point, however, research did not lead to satisfactory results. It was possible, indeed, to determine the chemical nature of certain of the extracted products, and even to obtain inductions with synthetic preparations of such substances. But a wide variety of organic compounds, such as fatty acids, nucleotides and sterols, had the same effect. The conclusion was inevitable that no specific chemical actions were involved, but that the ectoderm will react to the most diverse influences with the same reaction, namely, the formation of a neural plate. Experiments by Holtfreter (1945-47) made it likely that an injury to the host tissues is the primary effect in many cases of so-called induction by means of chemical substances, or of dead, or foreign tissue. Presumably, substances are thereby liberated which direct the differentiation of the cells into a course leading to a neural plate. This made it unlikely that experiments with such abnormal agents could be used to deepen our insight into the normal process.

In later years, therefore, another course was followed, which took Woerdeman's results (p. 113) as its starting point. Woerdeman found a very rapid decomposition of glycogen in cells of the marginal zone that were being rolled in. Heatley and Lindahl (1937) confirmed this, using quantitative microanalytical methods. Although it was found that glycogen itself could not be regarded as the evocator, yet attention was now called to the metabolic processes in the marginal zone material. Several investigations, in particular those of J. Brachet and of Barth and co-workers, showed that the cell metabolism in the organisation centre differs from that in the rest of the germ. Both the CO_2 production, and the respiratory quotient have higher values. Further research indicated that the main role is played here by protein metabolism, and not by carbohydrate metabolism. Attention was drawn to the sulphydryl compounds which are highly important in the oxidation and reduction processes in the cell, and later also to the ribonucleic acids, with which the sulphydryl compounds are closely linked. J. Brachet (1938, 1942) proved that in amphibian embryos these compounds are specially accumulated at places where the processes of development are most active. Originally, they are to be found in the nucleus of the oöcyte; later, they scatter over the egg cortex, especially at the dorsal side. During gastrulation, a considerable accumulation of ribonucleic acid and sulphydryl compounds occurs in the lips of the blastopore, and afterwards in the archenteron roof, in particular in those parts of the cells that face the overlying ectoderm. Brachet (1945) put forward the view that ribonucleic acid plays the leading role in induction. The inductive powers of dead inductors are proportional to their contents in ribonucleoproteins, and disappear on decomposition of these substances, e.g. by boiling or by enzyme treatment. However, later observations have shown that it is the integrity of the protein, rather than that of the ribonucleic acid, which is necessary for the activity of extracts and abnormal inductors. Several recent experiments point to the protein nature of the evocating substances from abnormal inductors. It must be kept in mind, however, that the mechanisms underlying induction by the living normal organiser might be completely different from those at work in these experiments.

Whatever their nature may in the end prove to be, we must in any case assume that the evocators originate in the cellular metabolism of the archenteron roof, and that they are taken up by the overlying ectoderm, and influence its chemodifferentiation so that it becomes able to form a neural plate. For this purpose, the ectoderm cells must be sensitive to the activity of these particular substances. We have seen that this is the case only during a short period. Therefore, the production of the evocator by the organiser, and the specific sensitivity of the reaction system must coincide. Presumably the sensitivity is due to the fact that at exactly the right moment the evocator fits the cellular mechanism of the reaction system, as a key fits the lock. It may be, for instance, that the evocator can catalyse certain reactions which take place at this stage in the cell's development, and which are impossible at other times. It is the mutual adjustment of the chemical activity of the cells of organiser and reaction system, that ensures the normal course of development.

We have already mentioned that the activity of a transplanted organiser is not restricted to the ectoderm of the host, but that the adjacent mesoderm is influenced as well. Evidently, the induction spreads from cell to cell in the latter (Fig. 41). This causes the cells of the host's mesoderm to differentiate into notochord and somite tissue. The inductor, therefore, makes this tissue similar to itself, hence the name *"assimilative induction"* for this process. As there are a number of differences between this type of induction and contact induction, we shall now discuss progressive induction.

We have already seen that the grafted organiser does not always form the same components of the secondary embryo. Often the materials of graft and host co-operate in the formation of its organs, giving them a "chimerical" composition, which varies from case to case, depending upon the experimental conditions. This is clearly illustrated by a number of experiments by B. Mayer (1935). He performed heteroplastic transplantations of a lateral half of the organisation centre of one embryo into the ventral side of another. One edge of the graft, therefore, consisted of mid-dorsal material which would normally have formed a notochord, and the rest consisted of prospective somites (cf. Fig. 34). The result of the experiment depended on the age of the gastrula from which the graft was taken. If it was taken from an older gastrula, its originally median part formed a notochord, whereas the remainder produced a single series of somites. Both areas, therefore, differentiated according to their original potencies (Pl. XIII, b).[1] Grafts taken from an early gastrula, however, produced a notochord flanked on both sides by somites (Pl. XIII, a).[1] This shows that "regulation" took place here, which made possible the development of a symmetrical archenteron roof from one half of the organiser, and led to a deviation in the differentiation of certain regions from their normal fate. In both cases, the host's mesoderm supplied the missing parts of the archenteron roof. The latter effect was also seen in another experiment, in which the graft formed only a neural primordium, whereas the host itself produced notochord and somites (Raven, 1938). The most important aspect of these facts is that, whatever the shares of graft and host in the composition of the organs may be, yet

---

[1] Facing page 138.

in many cases a more or less complete secondary embryo is formed. The tissues of graft and host supplement each other; one produces what the other cannot supply. There is endless variety, therefore, in the ways in which the secondary embryo may arise from the co-operation of graft and host. The use to which the cell material is put varies from case to case, and its origin from graft or host does not determine its fate. The organising process is not restricted to the organiser. From there, it extends into the surrounding tissues as well, and unites graft and host tissues into one integrated system. The differentiation of each cell depends only on its position in this system, and not on its origin. The best way to formulate these facts seems to be the statement that, starting from the organiser, an *organisation-field* arises.

The organisation-field is a system within which a number of developmental factors are distributed in such a way that the formation of certain organ primordia in the field is thereby determined. The fate of each element depends on its position in the field. The field possesses the power of *regulation*. This expresses itself in two phenomena. Firstly, it is not tied to definite material elements, but instead it displays a certain freedom as regards them. It can extend over the material, and readjust itself; the field can be "transposed". Secondly, its size may vary according to circumstances. After the removal, or addition, of material, it can adjust itself to the new conditions. We shall now discuss some examples of these regulative capacities.

Early gastrulae of *Triton* can be halved, by transection or constriction, in such a way that each part contains one half of the organisation centre. Two complete embryos will be formed in this case. Both may be slightly less well developed on the side of the transection, but in all other respects they are harmoniously built (Ruud and Spemann, 1922). Evidently an entirely new organisation-field has arisen in each half. The original field has been divided into two, and has, on a smaller scale, reconstructed itself in each half. However, if the early gastrula is halved frontally, so that its dorsal and ventral halves are separated, the ventral half, which has no organisation centre, will often produce only a "belly piece" lacking such axial organs as neural tube, notochord and somites.[1] The dorsal half, which contains the whole organisation centre, will nearly always produce a harmoniously built embryo of half the normal size (Fig. 43). Here, again, regulation takes place, for in spite of the fact that the whole organisation centre is present, the organisation-field is formed on a smaller scale, adapted to the lesser proportions of the halved embryo. Dalcq and Huang (1948) have repeated these experiments with a more accurate technique. They have shown that the processes underlying this regulation are of a far more radical nature than Ruud and Spemann suspected. An entirely new organisation

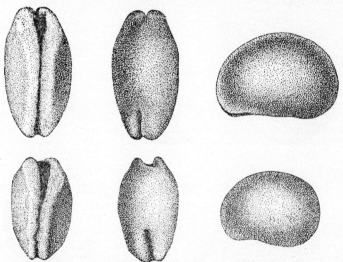

Fig. 43. Above: Closing neurula of **Triton taeniatus**, seen from dorsal, from ventral, and from the right side. Below: embryo developing from the dorsal half (after frontal transection of an early gastrula) at the same stage of development. The latter embryo is quite harmoniously built, as may be seen on comparison with the normal embryo. After Ruud and Spemann.

[1] According to later observations by Dalcq and Huang, however, such a ventral half will, if the ligature was made at the blastula stage, often form an embryo.

takes place in the halved germs, which disposes of the available material in a more or less "arbitrary" way, varying from case to case.

This power of regulation was also encountered in experiments in which material of the organisation centre was explanted, and cultured in a salt solution. In this case, it produced not only notochord and muscular tissue, but, if sufficient material was present (e.g. because several blastopore lips were cultured together), ectoderm would be formed as well (Lopashov, 1935). This ectoderm surrounded the other tissues, and here, too, in part of it the formation of nervous tissue was induced. In this way organ complexes were formed that had some faint re-semblance to embryos.

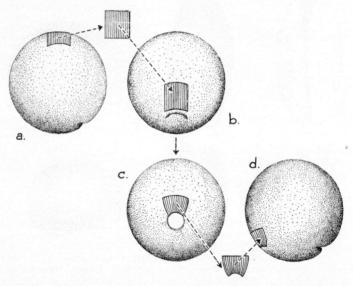

Fig. 44. Diagram of operation. (*a*) part of the presumptive ecto-derm of an early gastrula is (*b*) grafted into the dorsal lip of the blastopore of another germ of the same age. After 24 hours, (*c*), part of the graft has already been rolled in over the edge of the host's blastopore. The part that is not yet rolled in is now removed, and (*d*) grafted into the ventral side of a third gastrula.

The organisation-field can also extend over indifferent material that comes within its reach. Such material then acquires the power of organisation, and can thenceforward act as an organiser itself (Spemann and Geinitz, 1927). A piece of prospective ectoderm from one early gastrula, grafted into the dorsal lip of the blastopore of another, will in 24 hours acquire all the properties of blastopore lip material. If removed again, and grafted into the ventral side of a third embryo (Fig. 44), it will first of all prove to have acquired the topo-genetic capacity of independent invagi-nation, followed by differentiation into notochord and somites. Moreover, it has become an organiser which gives rise to the primordium of a secondary embryo at the place of implantation (Raven, 1938), (Fig. 45). This means that there we have another case of assimilative induction; the piece of ectoderm has acquired organiser properties by its stay in the organisation-field; this is the reason why it is able to give rise to a new organisation-field after transplantation into an indifferent area. It is possible, in other words, to transfer the organ-ising power to indifferent material. This is fresh evidence of the freedom of the field with regard to the material ele-ments of the germ.

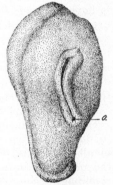

Fig. 45. The ecto-derm graft has ac-quired the properties of an organiser be-cause of its stay in the blastopore lip. It has organised the primordium of a se-condary embryo in the right flank of the host. *o*: secondary blastopore.

At first sight, it may seem that the concept "organisation-field" does not help us much to acquire a real under-standing of the phenomenon of organ-isation. It does not reduce the great problem, how qualitatively different organs can arise side by side from one common material. The concept, however, has shown its fecundity by giving rise to working hypotheses for the solution of this problem. Dalcq's hypothesis may be taken as an example. Its main features are that the organisation-field is supposed to be a gradient-field (cf. p. 47), due to local differences in the concentration of a substance, *"organisine"*, identical with that which is responsible for contact induction. Its concentration is thought to be highest in the centre of the dorsal marginal zone, and from there it decreases in all directions, though perhaps not everywhere at the same rate. This chemical gradient-field, in turn, is supposed to owe its origin to the interaction of the yolk gradient and the cortical gradient of the egg (cf. pp. 52 and 112). Here, however, it is the ratio, and not the product, of the factors C and V that is important (cf. Fig. 37). The origin of qualitatively different organs in an orderly spatial arrangement is assumed to be due to the dependence of cell differentiation upon the "organisine" concentration. At very high concentrations, a notochord is formed; at a lower concentration, somites, and so on. Which differentiation occurs, depends on whether or not the concentration exceeds a certain liminal value. The regulation phenomena can easily be explained on this basis. After disturbances in the germ, the organisine is redistributed. Simple diffusion leads to an equilibrium in which its concentration decreases again in all directions from a central maximum. The expansion of the organisation-field from a grafted organiser into the surrounding tissues can also be explained simply by diffusion of organisine, and the same applies to the transference of organising activity to indifferent material grafted into the organisation centre.

Later, Dalcq has extended his hypothesis by attributing great importance to the *"physiological competition"* between parts of the germ. This concept was introduced by Spiegelman (1945), who used it in a theoretical discussion of the phenomena of regeneration (cf. p. 181). Neighbouring groups of developing cells are supposed to compete by monopolizing food substances, and by flooding each other with the waste products of their metabolism. A cell group favoured in some way, would thus be able to suppress the differentiation of other groups in its environment. This is the so-called phenomenon of *dominance*. Dalcq (1947) applied this idea to the relationships in the organisation-field, and assumed that the cells compete for the limited quantity of "organisine" that is available. He succeeded, indeed, in giving simple explanations of several observed phenomena on this basis.

Dalcq's expositions formed a fertile working hypothesis which gave rise to extensive discussions and stimulated numerous further experiments. New data further arose from continued experiments on the inductive action of abnormal inductors and tissue extracts (cf. above, p. 125). It appeared that substances can be extracted from certain tissues which induce in a specific way distinct regions of the axial system, e.g. only the anterior brain region, or, on the other hand, structures of the posterior trunk and tail region. The results of these experiments seem to show that the "unitarian" theory, in which the regional differentiation both in the inducing archenteron roof and in the induced neural plate are explained by differences in concentration of one and the same substance, needs qualification. Presumably more inductor substances are involved in the process. Most authors now assume that at least two substances co-operate in the regional differentiation of the axial system.

Both are distributed according to a gradient-field. The first one, mostly called *"archencephalic inductor substance"*, presumably has its highest concentration in the anterior region of the archenteron roof. Acting alone, it induces in the ectoderm anterior brain parts with eyes, nasal organs, etc. A substance with similar action (but evidently not necessarily the same substance) is present in certain abnormal inductors, e.g. in guinea pig liver treated with alcohol. But even quite unspecific injurious influences, as treatment with acids or alkali, ammonia, alcohol, or distilled water, may evoke the same differentiations in gastrula ectoderm. Apparently, this first principle has a rather unspecific character; it is called *"activation"* by Nieuwkoop (1952).

The second substance variously called *"spinocaudal"* or *"mesodermal inductor substance"*, probably has its maximum concentration near the posterior end of the archenteron roof, and extends with decreasing concentration forwards up to the middle of the head region. Acting alone, it induces in the ectoderm the formation of mesodermal structures; in normal development, these are the posterior trunk and tail mesoderm, which arise from the posterior fifth of the neural plate (Spofford, 1948). A substance with similar action is found e.g. in guinea pig bone marrow treated with alcohol.

Regional differentiation, both in a craniocaudal and a mediolateral direction, now seems to be due to the interaction of these two agents. Differentiation in each area is dependent on the proportion of the two agents. According as the concentration of the mesodermal inductor increases relative to that of the archencephalic inductor, the differentiation changes from anterior to posterior brain parts, then to spinal cord and other trunk organs, then to tail structures, and finally to posterior tail mesoderm alone. This has been shown both by combining the relevant abnormal inductors (Yamada and Takata, 1955; Toivonen and Saxén, 1955) and in experiments with parts of the normal archenteron roof (Nieuwkoop and collaborators, since 1952, e.g. Sala, 1955).

Nieuwkoop (1952) has especially called attention to the time relationships between these two agents in normal development. According as the archenteron is invaginated during gastrulation, successive parts of the ectoderm come under the influence of the archencephalic inductor in the anterior part of the archenteron roof, and are "activated". With further invagination, each area of the future neural plate (except its most anterior part) then comes under the influence of the second (mesodermal) inductor, and its differentiation tendency to anterior brain parts, evoked by the activation, is "transformed" into a tendency to produce more caudal structures, the nature of which is dependent on the final strength of the transforming principle (cf. Eyal-Giladi, 1954).

Future experiments will have to show the chemical nature of the two inducing substances. Present results indicate that both the "archencephalic" and the "mesodermal" agent from abnormal inductors are proteins, but, as stated above, they are not necessarily the same substances that play a part in normal development.

CHAPTER IX

Induction and organisation

II. The period of organ development

Neurulation is immediately followed by a rapid increase in the spatial multiplicity of the embryo, caused by topogenetic processes which now begin at various places in the body. During gastrulation and neurulation, the topogenetic processes in the germ formed a single coherent system, but now the processes of folding, the cell migrations, and the invaginations and extrusions, are of a more local nature. They result in the almost simultaneous demarcation of a great number of *organ primordia*. We shall now discuss this period of organ formation or *organogenesis*.

There is conclusive evidence that here, again, a major role is played by the interactions of the parts of the germ which, under the name of induction and organisation, we have studied in the previous chapter. The development of the eyes in amphibians may serve as a first example for the discussion of these phenomena, because these organs were among the first to be studied from the point of view of developmental mechanics, and because analysis has already made very good progress here.

The vertebrate *eye* is a complicated organ. It arises by the combination of parts from different origins. After the neural plate has closed, thereby forming the neural tube, both lateral walls of the latter form a vesicular protrusion, the *primary eye-vesicle*, in the region of the prospective fore-brain (Fig. 46 a). These protrusions at first have a wide communication with the lumen of the neural tube, but this soon becomes constricted so that the eye-vesicles are connected with the brain only through the *eye-stalks*. The outer wall of the eye-vesicles soon establishes contact with the skin ectoderm covering the

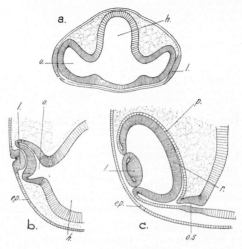

Fig. 46. Diagram of the development of a vertebrate eye. (a) protrusion of the eye-vesicles (o) from the brain (h); l = lens primordium; (b) invagination of the eye-vesicle, resulting in the formation of the eye-cup; formation of the lens-furrow, (l); (c) eye-cup still connected with the brain through the narrow eye-stalk (o.s.); differentiation of the wall of the eye-cup into retina (r) and pigment epithelium (p); lens (l) has become free from the epidermis (ep).
After Kühn.

sides of the head. At this point, the ectoderm thickens, thereby forming the *lens-placode* which then becomes invaginated, and, as a *lens-vesicle*, disengages itself from the ectoderm. At the same time, the eye-vesicle also invaginates, thereby forming the double walled *eye-cup* (Fig. 46 b). The inner wall of the cup becomes thicker, and develops into the *retina*; the outer wall remains thin and forms the *pigment epithelium* surrounding the retina (Fig. 46 c). Later the *iris* will arise from the edges of the eye-cup, and the eye-stalk will develop into the *optic nerve* which connects the eye with the brain. The transparent *cornea* is formed by the skin overlying the eye. Finally, the mesenchyme surrounding the eye also contributes its share. It penetrates into the eye-cup and takes part in the formation of the *vitreous body,* and it also forms several tough membranes around the eye.

One of the first problems arising in connection with the development of the eye is: What causes the protrusion of the eye-vesicles from the wall of the brain? Experiments in which

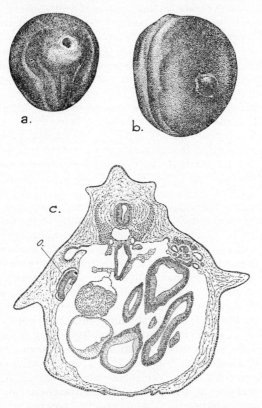

Fig. 47. (*a*) the right eye primordium is removed from the rostral part of the neural plate of one embryo of **Bombinator**, and (*b*) is grafted into the right side of another embryo; (*c*) cross section of the larva developed from the embryo pictured in (*b*); an eye-cup (*o*) has developed from the graft. After Spemann.

parts of the rostral end of the neural plate were transplanted into other areas of the germ have shown that such grafts could develop into normal eyes (Fig. 47). Evidently, the potency for the formation of an eye-vesicle is already present in the anterior parts of the newly formed neural plate. Mangold (1928-29) has even shown that the material concerned possesses this eye-forming potency as early as the late gastrula stage, before there is any neural plate at all. The ectoderm acquires this potency under the influence of induction by the anterior part of the archenteron roof. Experiments by Spemann (1931) have shown that, after transplantation into the ventral side of another embryo, this part of the organisation centre induces not only brain tissue, but eyes as well (Pl. XII, b). It is noteworthy that the eye-forming potency is at first evenly distributed throughout the anterior region of the neural plate, so that any part of it will form an eye after transplantation. It is not until later that the eye-forming power becomes concentrated in two lateral centres, and lost in the median region. Experiments by Adelmann (1929-37) have shown that this, too, is due to the activity of the archenteron roof. From a certain stage onwards, a narrow median strip of the archenteron roof begins to exert an induction which suppresses the eye-forming potency in the overlying ectoderm. This observation supplies the explanation of a common deformity: chemical or mechanical injury to the median part of the archenteron roof in the head region destroys its inhibitory influence, and therefore, instead of two lateral eyes, a single median eye now arises from the bottom of the neural tube (so-called *cyclopia*).

The next problem is whether, simultaneously with the protrusion of the eye-vesicle, its later differentiation into retina, pigment epithelium, and eye-stalk is also unequivocally determined. A grafted eye-primordium invaginates in the normal manner, and forms an eye-cup. The differentiation of retina and pigment epithelium also takes its normal course. This shows that the determining influence of the archenteron roof has also to some extent fixed the later phases of the development of the eye-cup. Experiments by Dragomirow (1932-37), however, have shown that, at the eye-vesicle stage, parts of the

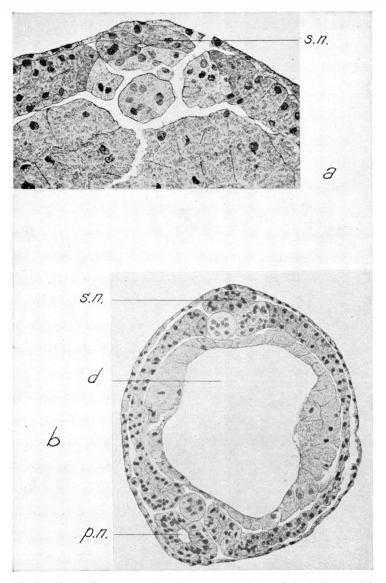

PLATE XIII. The co-operation between organiser and host tissues. (a) Secondary embryo primordium, organised by one half of the blastopore lip of an early gastrula. The graft (light pigmentation) produced the notochord, flanked on either side by somite material, and the floor of the secondary neural tube (s.n.). The host produced all the remainder. (b) Secondary embryo primordium, organised by one half of the blastopore lip of a late gastrula. The graft has formed only the notochord, and the somites of one side. The host has produced all the rest. p.n. = primary neural tube; s.n. = secondary neural tube; d. = lumen of the common gut. After Mayer.

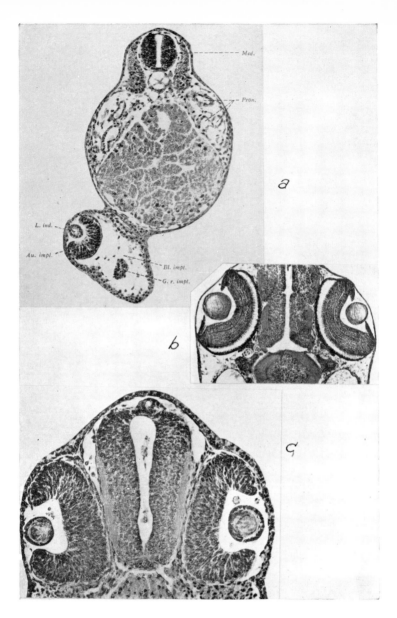

PLATE XIV. Induction of a lens. (a) Lens induction (**L. ind.**) in the belly epidermis of **Triton taeniatus** by an implanted eye-cup (**Au. impl.**) of **T. alpestris**, after Mangold. (b) The eye-cup of **T. taeniatus** has induced a lens in transplanted ectoderm from **T. cristatus** (left). This lens is larger than a normal **taeniatus** lens (right). (c) The eye-cup of **T. cristatus** has induced a lens in transplanted ectoderm from **T. taeniatus** (left). This lens is smaller than a normal **cristatus** lens (right). (b-c) after Rotmann.

vesicle wall can still replace each other. The final determination of the lateral wall to form a retina does not take place until the beginning of the invagination of the vesicle. Later experiments have shown that the appearance of the retina depends on the aggregation of the cells of the eye vesicle to a massive multilayered structure. Normally this occurs predominantly at the site of contact with the ectoderm; in experimental cases it can also occur at places where contact is established with other epithelial organs, such as the ear vesicle. On the other hand, pigment epithelium forms where the cells are stretched into a single layer, and are covered on their outer side by mesenchyme cells (Lopashov, 1961).

Let us now turn to the development of the lens. Transplantation experiments have proved that in some amphibians the ectoderm that will later form the lens has a weak lens-forming potency already at the late neurula stage. If it is grafted, e.g. into the ventral side, it will produce more or less well developed small lenses (Harrison, 1920; Woerdeman, 1934). This early lens-forming potency is apparently due to inductive actions from the underlying archenteron roof (Jacobson, 1955). It becomes much stronger at the time when the protruding eye-vesicle comes into contact with the ectoderm. The assumption was obvious that an inductive action from the eye-vesicle on the overlying ectoderm plays a role here. The following experiments proved that this hypothesis was correct. (1) In many amphibians, no lens is formed if the eye-vesicle is removed before it establishes contact with the ectoderm; in other species, only a slight thickening of the epidermis will occur, and in only a few (e.g., *Rana esculenta*) a well developed lens is formed, apparently independently of the eye-vesicle (Spemann, 1912), (Fig. 48). (2) If the lens ectoderm is replaced by other ectoderm, originating, e.g., from head or belly, a lens will under certain circumstances be formed by this foreign ectoderm. (3) The same thing happens if an eye-vesicle is cut out, and grafted under the ectoderm of another part of the body (Lewis, 1904), (Pl. XIV a).

However, in the last two experiments, it appeared that the

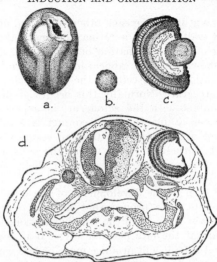

Fig. 48. Independent differentiation of the lens in **Rana esculenta**. (*a*) extirpation of the right eye primordium at the neural plate stage; (*d*) cross section of the resulting larva, 14 days after the operation; the right eye (left in this figure) is absent; nevertheless, a lens (*l*) has been formed, which, however, (*b*) is smaller than that of (*c*) the normal left eye. After Spemann.

reaction of the ectoderm is not always the same in this respect. At early stages, it is true, the whole ectoderm still has the power to react to induction from the eye-vesicle with the formation of a lens. But at least in some amphibians, after neurulation this capacity becomes restricted to the more or less immediate neighbourhood of the eye. The area of the "re-action-field" decreases in the course of development. The capacity to react, the *competence*, remains strongest at the place where the lens is normally formed, and decreases gradually from this centre toward the periphery.

Some details of this induction process may be mentioned. First of all, it appears that there is a correlation between the size of the induced lens and that of the inducing eye-vesicle. The size of the latter can be reduced or increased experimentally, by extirpation of material or fusion of two eye-primordia, respectively. Such modified eye-vesicles always induce a lens which approximately fits the eye-cup (Fig. 49). But this is only true if eye-vesicle and lens ectoderm belong to the same species, and not in the case of heteroplastic combinations. Rotmann (1939) has shown that under the influence of the (larger) eye-vesicle of *Triton cristatus* the ectoderm of *T. taeniatus* forms a lens of the normal *taeniatus* size, and therefore too small in comparison with the eye. In the reverse combination, the *cristatus* lens is too big for the *taeniatus* eye-cup (Pl. XIV b-c).

These relationships have been further studied by Balinsky (1957a), in experiments in which eight species of frogs and toads were used. He showed that the size of the lens rudiment is dependent both on the size of the eye cup, and on the reactivity of the epidermis. The latter showed great differences from species to species, in contradistinction to the power of induction of the retina, which appeared to be the same in all species investigated. Moreover, there is an upper limit for the size of the lens in each species, and the increase in the volume of the eye cup becomes less and less effective in enlarging the lens rudiment, the nearer the latter is to this limit. Therefore, the dependence of the size of the lens rudiment on the size of the eye cup has to be expressed in the form of an exponential equation.

No continuous activity of the eye-cup is necessary for the final differentiation of the lens. If an eye-vesicle is removed after it has influenced the ectoderm for

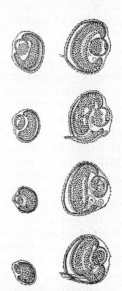

Fig. 49. The size of the lens depends on that of the eye-cup. Left: small eye-cups of **Bombinator**, obtained by the removal of varying quantities of material. Right: the normal eye-cups of the other half of the body. After **Spemann**.

24 hours, the lens primordium, as yet consisting only of a slight thickening of the ectoderm, will go on differentiating independently (Filatow, 1934). The internal wall of the lens-vesicle always supplies the *lens fibres,* and its external wall the *lens epithelium.* This polarity of the lens, too, is determined by the eye-cup. However, other organs than the eye-cup can exert the same influence. If the lens-vesicle lies in an abnormal part of the head, and is in contact with an ear-vesicle, an olfactory pit, or with the brain, lens fibres are formed at the place of contact (Balinsky, 1930; Dragomirow). Apart from its polarity, the lens also has a bilateral structure. Woerdeman (1939-41) has shown that the various processes which play a part in the normal development of the lens, such as the formation of the lens-placode, the invagination of the lens-vesicle, fibre differentiation, and the development of polarity and bilateral symmetry are determined not simultaneously, but successively.

The induction which gives rise to a lens is not species-specific. Not only in heteroplastic, but even in "xenoplastic" transplantations a lens may be induced, e.g., by an anuran eye-vesicle in the ectoderm of a urodele, and *vice versa* (Holtfreter, 1935).

Evidence was obtained by means of serological methods that specific lens proteins are already present in the lens-primordium at the lens-groove and lens-vesicle stages, i.e. before the beginning of actual fibre differentiation (Ten Cate and Van Doorenmalen, 1950). Furthermore, Woerdeman (1950) demonstrated that such lens proteins are also formed in mixtures of extracts of eye-vesicles and of head-ectoderm from early embryos in which no lens formation has yet begun. With further development of the lens gradually more antigens are formed. Langman showed that anti-lens sera have a noxious influence on the embryonic chicken lens (Woerdeman, 1962).

A further group of experiments was made on the origin of the cornea. Both the eye-cup and the lens may induce the formation of a cornea. In normal development, therefore, this tissue is apparently formed under the combined influence of both factors (Lewis, 1905; Fischel, 1915-19). The whole skin ectoderm of the larva has the capacity to differentiate into a

cornea under the influence of these inductions. The presence
of the eye is necessary, not only for the development of a
cornea, but for its maintenance as well. After the extirpation
of eye-cup and lens, de-differentiation of the cornea takes place,
and it alters into normal skin. Here, again, the induction is not
species-specific.

It is clear, therefore, that the normal development of the
eye is based on an interplay of inductions, whereby each part
influences the development of all others, and also reacts to the
influences from all others, all parts having acquired the necessary reactivity in their previous chemodifferentiation.

Similar situations obtain in other organs. The *olfactory organ*
of the amphibians originates from a pair of thickenings of the
ectoderm in the anterior region of the head. Here, in front of
the eye-vesicles, the lateral walls of the neural tube come into
very close contact with the ectoderm. Later this part of the
brain will form a paired protrusion, the *cerebral hemispheres.*
The ectoderm which is in contact with these parts of the brain
wall thickens into a pair of *olfactory placodes,* which later, by
invagination, form the *olfactory pits.* Several experiments
have shown that this differentiation of the ectoderm is also
caused by inductions from the brain wall (Raven, 1933; Haggis,
1956). If the rostral end of the brain is double, either partly, or
completely, an olfactory pit is formed in contact with each of
the cerebral hemispheres. If one hemisphere is absent, the corresponding olfactory pit is absent, too. In cyclopic embryos, the
fore-brain is poorly developed because of deficiencies in the
anterior part of the archenteron roof. In such cases, there is
not only a single median eye, but also a single olfactory pit,
situated in front of the anterior end of the brain. This, again,
is a non-specific induction. Holtfreter (1935) found that, in the
xenoplastic combination of anuran with urodele cell material.
induction of olfactory pits was still possible.

In the case of the olfactory organs, the stage at which the
definitive determination by contact induction takes place is
probably preceded by a phase during which there is a slight,
labile "predetermination" already, just as we have seen in the
case of the lens of the eye (above, p. 139). In particular in some

anurans this early phase of determination may be very important, and can replace induction by the fore-brain completely, or nearly so (Zwilling, 1940); it probably originates under the influence of the archenteron roof.

The *auditory organ,* too, arises from an ectodermal thickening, the *ear-placode,* which invaginates and becomes separated from the ectoderm, forming the *ear-vesicle.* This gives rise to the various parts of the internal ear by a series of folding processes. In urodeles the area concerned generally acquires an autonomous capacity to form ear-vesicle during the closing of the neural tube. In anurans this may happen a little earlier. Yntema (1939), however, has shown that in *Amblystoma punctatum* a first, weak determination takes place at a much earlier stage, viz. at the late gastrula stage.

These phenomena were further investigated by Harrison (1936). Initially, the potency to form ear-vesicle is not strictly limited to the material that will later actually produce it. It extends also over the surrounding ectoderm, though there is a gradual decrease in intensity. Any cell group from this zone can produce a complete ear-vesicle after transplantation, e.g. into the ventral side. The size of such an ear-vesicle depends upon that of the graft, but its structure may be entirely normal. Several ear-vesicles can originate from one "prospective ear region" in this way, and, conversely, two such regions can fuse if grafted side by side. In the latter case, they produce one ear-vesicle of double size. This proves that here, too, we find an organisation-field possessing a certain freedom with respect to the cell material.

Harrison has also investigated how the symmetry of the auditory organ is determined. Each of these organs is in itself asymmetrical, for it has different shapes in the three dimensions. But one auditory organ is symmetrical to that of the other side of the body. It is an important problem how the determination into a right or left-side organ takes place. As we shall discuss this more fully in connection with the limbs, the following remarks will suffice here. Initially, the ear primordium is still unpolarised, so that after transplantation it develops into a left or a right ear in accordance with its new environment. But soon the rostrocaudal axis, and then also the dorsoventral axis, become fixed. Once this has happened, these axes cannot be reversed any more, and when inverted the primordium will go on developing in accordance with its original polarity. From then on, a left primordium will always become a left auditory organ, even if transplanted into the right side.

At a still later stage the determination of the component parts within the "ear-field" takes place. After this, a grafted part of the field no longer produces a complete auditory organ, but only certain parts of one. Exchange of cell groups now results in ear-vesicles of abnormal composition (Kaan, 1926). This shows that the ear-field is no longer a unity, but that it is broken up into a mosaic of parts with different potencies. Moreover, the ear-forming potency is now restricted to those cells that will really take part in the formation of the ear-vesicle; the peripheral parts of the ear-field have lost this potency.

For a long time it remained doubtful what, in normal development, induces the ear-vesicle. On the one hand, Dalcq was able to show that in *Discoglossus* there is a centre of ear-vesicle induction in the lateral parts of the archenteron roof. On the other hand, Trampusch (1941) made it seem likely that an important role is played in the induction of the auditory organ by a cell group lying in the neural folds during the neural plate stage. Later, this group, the *neural crest*, migrates in a ventral direction under the ectoderm at the sides of the head. Extensive experiments by Yntema (1950) have elucidated this problem, at least for *Amblystoma punctatum*. According to his results, the determination of the auditory organ takes place in two phases. At the late gastrula or early neurula stage, a first, weak determination is effected by the archenteron roof. From the late neurula stage onwards, a second induction takes place, this time by the brain or the neural crest. This gives the determination of the ear-vesicle a more final character. There are qualitative differences between these two inductions, and in the ectoderm two qualitatively different phases of competence correspond to them. In other words, each induction attains its maximum at about the same time at which the ectoderm's ability to react also reaches its corresponding peak.

Finally we must mention that xenoplastic induction of the auditory organ between anurans and urodeles is possible (Holtfreter, 1935). Such a xenoplastic labyrinth develops normally for a long time. The speed of its development, and its size and proportions are entirely analogous to those in the species to which the ear-forming ectoderm belongs. It may even function normally in the foreign larva. At a certain stage in larval life, however, an incompatibility reaction suddenly sets in, and the xenoplastic labyrinth rapidly disintegrates (Andres, 1949).

The anterior end of the archenteron, which was the first to invaginate during gastrulation, comes into close contact with the rostral ectoderm. A *mouth plate*, consisting of ectoderm and endoderm is thereby formed. Later this breaks through, and a *mouth* is formed. In urodeles this process is due to induction by the endoderm of the archenteron wall. If, at the neural plate stage, this endoderm is grafted somewhere under the ectoderm of the flank, a mouth cavity will be formed at this point (Ströer, 1933). Balinsky (1948) has demonstrated that here, too, a weak predetermination precedes the final determination of the mouth. Probably this predetermination takes place during gastrulation. Recent experiments have shown that in anurans the relationships are somewhat more complex. Here the neural crest arising from the anterior transverse neural fold also plays a part in mouth induction, together with the endoderm of the mouth region. The results of xenoplastic transplantations are highly interesting. In urodeles (newts), the larvae have real teeth, consisting of dentine. In the anurans (tail-less amphibians), however, they have horny teeth, formed by local proliferation and cornification of the mouth epithelium, and the mouth edges are covered with horny jaws. Spemann and Schotté (1932) grafted belly ectoderm of anurans into the ventral side of the head of *Triton*. This ectoderm reacted to the inductive activity of the *Triton* mouth endoderm, with which it came into contact, by forming a mouth cavity. This, however, had the character of an anuran mouth, and was provided with horny jaws and teeth. Moreover, immediately behind the mouth, a pair of adhesive organs, consisting of glandular epithelium, developed from the grafted anuran ectoderm. These organs are always found in the corresponding place in anuran larvae, but the urodeles do not have them (Pl. XV). The tissue of the *Triton* host, therefore, had induced organs in the graft which the host itself did not possess. Holtfreter (1935) made the reverse experiment, which had the expected result. Belly ectoderm of *Triton*, grafted into the head of an anuran embryo, produced a mouth with dentine teeth. In these cases chimeric tooth germs may be formed, in which the odontoblasts are derived from the anuran host, although the latter in normal development does not possess any true teeth in the larval stage (Wagner, 1955; Henzen, 1957).

These experiments give a very clear illustration of the nature of induction. The cells of the graft react to an "order" given by the inductor, but they do it in their own way, by virtue of their own species- and organ-specific reactive powers, which are due to their genetic constitution and their previous chemodifferentiation. Induction does nothing but activate a definite developmental capacity, present in the cell, and dependent on the nature of the latter. Experiments by Balinsky (1955) have shown that this conclusion especially holds with respect to histogenetic processes, concerned with the differentiation of special types of cells and with their mutual arrangement e.g. presence or absence of horny jaws and horny teeth, and shape of adhesive organs. On the other hand, organogenetic processes, involving the number or mass of cells and the position of areas differentiating in a specific way (e.g. size of the mouth or the adhesive organ), may be partly or entirely dependent on the host. Apparently, differences in organogenetic processes between related species may be due to a modification of the inducing systems.

Somewhat more caudally, a row of lateral extrusions, the *branchial pouches,* is formed on each side by the walls of the gut. These endodermal pouches come into contact with the ectoderm, and at these points *branchial grooves* are formed by the ectoderm. Later the grooves break through into the pouches,

thereby forming the *gill slits*. The *branchial arches* separating
the slits are lined with endoderm on the inner side. Externally,
they are initially covered only by the ectoderm, though later
the endodermal epithelium grows out under this ectoderm.
Within each branchial arch there is a band of mesoderm, which
produces the muscles, and some mesenchyme originating from
the neural crest, which forms the cartilaginous skeleton of the
branchial arches. Branched *external gills* grow out from the
outer surfaces of the hindmost three pairs of branchial arches.
They consist of epithelium, filled with mesenchyme.

The formation of these gills was found to be due to induction
by the endoderm, the former archenteron wall. If gill region
endoderm was grafted under the ectoderm of the flank, gills
were formed there (Severinghaus, 1930). Here, too, it appears
that the gill-forming potency initially extends over a large area,
its intensity decreasing from a central maximum toward the
periphery. As long as the graft is not too small, a normal
set of gills develops from any part of this *"gill-field"* on transplantation. Two gill-fields grafted side by side fuse into one,
and produce one normal set of gills, provided that both were
oriented in the same way (Harrison, 1921). The gill endoderm
is not only responsible for the origin of the gills, but it also
determines their polarity. This was proved by inverting the
endoderm, when the resulting gills also were inverted. Moreover, the endoderm determines the specific form, size and
pigmentation of the outgrowing gills. This was proved by
heteroplastic transplantations between species of the genus
Amblystoma (Harrison, 1927). The ectoderm, however, is not
altogether without influence on gill development. Rotmann
(1931-35) exchanged the gill-region ectoderm between *Triton
taeniatus* and *cristatus,* and between axolotl and *Triton.* He
found that the gill-ectoderm determines especially the speed
of gill development during the earlier stages (Fig. 50). It was
also shown that it influences the size of the gills.

Very many experiments have been made in connection with
the *limbs* of amphibians. Their first visible primordia are
formed by the local accumulation, under the ectoderm of the
flank, of mesenchyme originating from the lateral mesoderm.

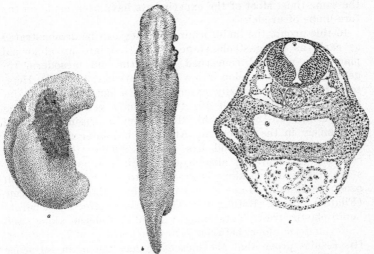

Fig. 50. (a) Embryo of **Triton cristatus**, into which a piece of prospective ectoderm of **T. taeniatus** has been grafted at the gastrula stage (cf. Fig. 38b). The graft occupies the prospective gill area of the right side. (b) Dorsal view of the same embryo at a later stage; development has proceeded farther in the gills on the right, formed by the **taeniatus** ectoderm, than in those on the left (formed by the host). (c) Cross section through the gill region of the embryo (b) (the gill primordium formed by the graft is on the left in this figure). After Spemann.

First, a half-spherical *limb-bud* is formed. This then assumes a conical shape, and subsequently begins to expand at its tip, thereby forming the *hand-* (or *foot-*) *plate*. The digits are formed by indentation of the edge of this plate, and become elongated in a certain sequence. Still later the joints are formed, and the limb is rotated so that it assumes the normal position relative to the body. Meanwhile, the skeleton and the musculature of the limb have differentiated from the mesenchyme, and the girdle skeleton with its muscles has developed in the flank. In urodeles, the fore-limbs are formed at an early stage, whereas the hind-limbs develop much later. In anurans, on the other hand, both pairs are formed at about the same time. Most of the experiments have been made on the fore-limbs of urodeles.

In this group, the limb-forming potency can be demonstrated as early as the gastrula stage. If grafted into an abnormal place, the material concerned (ectoderm and mesoderm together) will still develop into a limb (Detwiler, 1929-33). Here, too, the potency initially extends over a larger area, its intensity decreasing in all directions from a central maximum (Harrison, 1915-18). Probably the potency is originally located exclusively in the mesoderm. At somewhat later stages, however, when the limb-bud has become visible, the ectoderm covering the bud has also acquired this potency. If at this stage this ectoderm alone is grafted into another place, it will produce a limb in co-operation with the local flank mesenchyme (Filatow, 1930). Rotmann (1931-33) has effected separate heteroplastic transplantations of limb mesoderm alone, and limb ectoderm alone, between *Triton taeniatus* and *T. cristatus*. His results prove that the mesoderm has the main influence on the limb's specific form and size. It is not until later larval stages that a slight influence of the ectoderm on the size of the limb and the shape of the digits becomes manifest.

At first, the limb rudiment has all the properties of an organisation-field: transplanted parts of the primordium can produce a complete, harmoniously built limb, so that one primordium can give rise to as many as four limbs. On the other hand, two fused primordia can produce a single harmonious limb. All the cells of the rudiment, therefore, are still able to produce any part of the limb. The field has not yet become tied to certain cells, but it can be displaced, can divide into a number of equivalent fields, or can fuse with another field into one unity. The field can be "transposed" with regard to the material.

Harrison (1915-25) carried out a large number of experiments on the symmetry relationships of the limbs. Left and right limbs are mutually symmetrical. Here, again, we face the problem of how the development of a primordium into either a right or a left limb is determined. In order to solve this, Harrison transplanted limb rudiments of *Amblystoma punctatum*

in various stages, rotating them 180° on any one of their three
axes (viz. rostro-caudal, dorso-ventral, and medio-lateral), or
on two or three axes simulta-
neously. If these experiments
were made at an early stage,
a short time after neurulation, Harrison found that the
dorso-ventral and the medio-
lateral organisation of the
growing limb were always
in harmony with that of its
new environment. If, how-
ever, the limb had been
rotated on its rostro-caudal
axis, it grew out obliquely
forwards, instead of backwards, so that the symmetry
was reversed (Fig. 51). This

Fig. 51. **Amblystoma** larva, into
which a right limb bud, rotated
on its rostrocaudal and dorso-
ventral axes, was grafted in the
flank. The graft (*tr*) has produced a limb with normal dorso-
ventral structure, in which, however, the rostrocaudal organisation
is inverted so that the graft is
the mirror image of the normal
right fore-limb. After Harrison.

shows that, at the time of the operation, the rostro-caudal
axis was already irrevocably fixed in the material, but that
the other axes had not yet been determined. Further experiments showed that the determination of the rostro-caudal axis
takes place at the gastrula stage already, whereas that of the
other axes does not occur till later stages, irreversible determination taking place first in the dorso-ventral, and then in the
medio-lateral axis. Moreover, it was found that the different
species of urodeles did not behave identically in this respect.
In *Triton taeniatus,* for example, the direction of the dorso-
ventral axis is determined much earlier than in *Amblystoma*
(Brandt, 1922-28). The sequence, however, in which the three
axes become determined, is probably the same everywhere.
Presumably, the limb mesoderm is solely responsible for this
polarity, and the ectoderm does not play a role on this point
(Balinsky, 1931).

These results obtained in urodeles cannot be generalised.
Experiments by Zwilling (1955-56) have shown that the limb
ectoderm in the chick plays a more important part in development. When the limb ectoderm is rotated 90°, the axial

relationships develop in conformity with the ectodermal component. Apparently, a thickened apical ectodermal cap or ridge is especially important for the outgrowth of the limb. It induces the distal growth of the underlying mesoderm. However, the mesoderm seems also in birds to be the most important of the two. It determines the quality of the developing limb. Mesoderm of the hind-limb bud, covered by ectoderm of a fore-limb bud, develops into a hind-limb, and inversely. Hind-limb mesoderm of a duck, covered by ectoderm of a chick, develops into a leg with webbed toes, like a duck's (Hampé, 1959). Moreover, it has been shown that mesoderm of an early limb-bud, placed in contact with ectoderm of the flank, is able to form an extra limb (Kieny, 1960).

Interesting results were also obtained in successful attempts to induce limbs in abnormal places by grafting material into the side of the body. Balinsky (1925-27) discovered that in *Triton taeniatus* grafting of an ear-vesicle under the ectoderm of the flank often results in the outgrowth of a limb in this place. That this is not a *specific* induction is shown clearly by the fact that Balinsky obtained the same result by grafting an olfactory pit, or even a piece of celloidin, under the flank ectoderm. He made an extensive analysis of this phenomenon in a series of further experiments (1929-37). He found that a limb can be induced anywhere in the flank. The percentage of successful inductions decreases from the fore-limb region in a caudal direction, but increases again in the vicinity of the hind-limb. In the anterior regions of the body, the induced limbs have the character of fore-limbs, farther caudally more that of hind-limbs. The more caudal the induced limb, the later it will develop, irrespective of the moment of implantation. At a certain stage, however, the capacity to react to this induction disappears. Probably the whole of the flank of a urodele has a latent limb-forming potency in such a way that there is some sort of an equilibrium between the tendencies to form "body-wall" and "limb". Presumably, a non specific agent may lead to preponderance of the latter tendency. Originally Balinsky was inclined to think that this was caused by a local increase in the intensity of tissue metabolism. However, later observations using cytochemical methods and radioactive tracers failed to support this assumption. But it was found that the basement membrane of the epidermis is completely absent in the induced limb rudiments. Perhaps the retarding action of the graft on the development of this basement membrane is the primary factor in limb induction (Balinsky, 1957b).

A large number of experiments might be added to those mentioned above. These, however, will suffice to give an impression of the very complicated interplay of inductions that is at work in the embryo during this phase of development. It is hardly an overstatement to say that all parts of the embryo which are in contact, or which in the course of topogenesis come into contact, influence each other by contact induction. Each part, therefore, is at the same time an inductor, emitting certain influences, and a reaction system, reacting to the influences received. The physicochemical constitution, and therefore the potency, of each cell or cell group is modified according to the inductions it encounters, either simultaneously or successively.

It is highly probable that this contact induction is brought about by material influences, i.e. that the cells secrete substances which diffuse into the environment, and which are taken up by, or at least act on, the neighbouring cells. It must be admitted that for most inductions this has not been proved with the same certainty as for neural plate induction. In many cases, however, there are plenty of indications which support this assumption. The fact that dead tissues, and tissue extracts, often can induce not only neural tissue but many other organs as well, is one strong argument in its favour.

We must now turn our attention to the problem of how far the specificity of the effect depends on the nature of the induction, i.e. whether each organ is induced by its "own" evocator. Several possibilities must be considered here. One and the same evocator might be operating in all cases. The divergent development of the cell groups, e.g. into organs so different as an eye-lens, an olfactory pit, or an auditory organ, would then be due to differences in the reactivity of these cell groups themselves. The influence of induction would then be of a purely *activating* nature; it would not cause the intrinsic differences between the parts of the germ, but only make them manifest. We must not forget here that such a specific re-activity of the cell groups must owe its origin to previous chemodifferentiation. This, in its turn, would require an explanation.

Another attempt at an explanation also takes one evocator as its starting point, but attributes great importance to varia-tions in the *concentration* of this substance. The occurrence of specifically different organ-forming potencies would, in this view, be a consequence of concentration differences, to which the reaction system would be extremely sensitive. Which of two possible courses development would take would depend on whether or not the evocator concentration exceeded a certain "threshold". We have already seen (p. 132) how Dalcq ex-plained the properties of the primary organisation-field of the germ by means of this hypothesis.

But even in that case this "unitarian" view did not suffice to explain all phenomena of determination, and had to be replaced by the hypothesis of the existence of two interacting substances, each distributed according to a gradient-field. Local variations in the absolute concentrations of these substances, and of their mutual proportions, may create a multitude of different conditions arranged in a definite pattern. If we admit, as a third variable, the existence of possible variations in the reactivity of the cells, the divergent development of neigh-bouring cell groups may be easily understood. This may also account for the fact that even dead tissues and tissue extracts may induce organ complexes of different composition, but often showing a certain resemblance to the normal topography of distinct body regions, e.g. fore-brain, eyes, and olfactory pits, or spinal cord, trunk musculature and kidney. For the sub-stances emanating from such abnormal inductors will also tend to decrease along a concentration gradient, simulating in this way the distribution of the "normal" inductor substances.

It appears, therefore, that the determination processes at early stages (gastrulation and early neurulation) may be ex-plained by the interaction of only a few inductor substances.

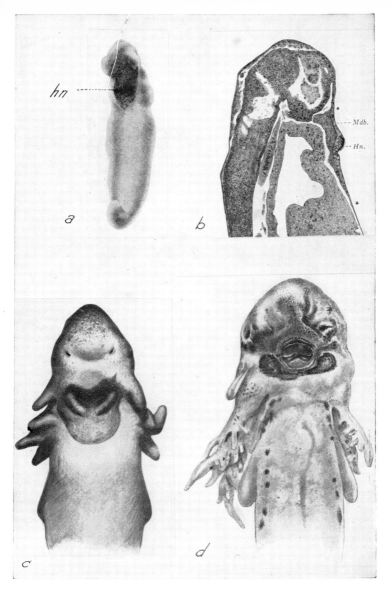

PLATE XV. The specificity of the reaction to the stimulus of induction. Belly ectoderm from a frog was grafted into the mouth area of a newt embryo. Here (**b, d**) it forms a mouth with horny teeth and a pair of adhesive organs (**Hn**). Both organs are characteristic of *f*rog larvae, and do not occur in newts. After Spemann and Schotté.

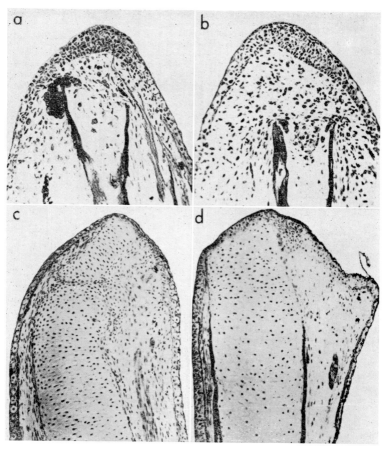

PLATE XVI. The role of an apical epidermal cap in limb regeneration. (a) Tip of amputated salamander limb after 4 days of regeneration. Apical cap and beginning accumulation of blastema cells. (b) 5-day regenerate. Apical cap and early blastema. (c-d) Amputated limbs, in which the formation of an apical epidermal cap was prevented by means of ultraviolet light irradiation. (c) 2 days after amputation. (d) 8 days after amputation. No apical cap and no blastema. (a-b) after Thornton (1956); (c-d) after C. S. Thornton, **J. Exp. Zool. 137**, 1958.

On the other hand, it seems probable that in the course of further development, with the progress of chemodifferentiation, more and more qualitative differences arise in the chemical constitution of the evocators for various organs.

Special attention must in this connection be paid to the situation in the head-region. We have seen above that we know, for several organs of the head, e.g. lens, olfactory pits, auditory organs, mouth, that their determination takes place in two phases: (1) an early phase, at the end of gastrulation, during which a weak tendency to form these organs arises in the areas concerned, and (2) a later phase, during which specific contact inductors, e.g. retina, hemispheres of the brain, neural crest, mouth endoderm, bring about the final determination of the organs. The first phase, which we have called "predetermination", apparently takes place under the influence of the archenteron roof, but it is probable that an even earlier gradient-field in the ectoderm also plays a role (Nieuwkoop, 1947; Lombard, 1952). In any case, there are strong indications that this predetermination of the organs of the head takes place, in a common "head organisation-field", in the way discussed above (p. 132). This view is supported, for instance, by the fact that treatment of the germ with lithium ions modifies the pattern of the head as a whole. Probably mainly quantitative differences in the concentrations of the evocators play a role here. In the second phase, the place and character of the organs would be irrevocably determined by qualitatively different contact inductions.

The evocators probably originate as products of the cell metabolism of the inductors, and diffuse into the environment. If they are to exert an effect on the reaction system, the cells of the latter must be sensitive to the action of just these substances. This is often the case only during a very limited period. Here, again, the chemical processes in the cells of inductor and reaction system must be accurately adjusted to each other for normal development to be possible. If the evocator is not produced at the right moment, or if the reaction system has not yet acquired its sensitivity by that time, or has lost it again, development will be upset. Whether or not these developmental processes will be interlinked in the correct way depends on the course of events in the previous phases of development. This explains why a slight disturbance at a certain moment may give rise to serious abnormalities in later stages.

The inductions cause the formation of the primordia of the various organs. These primordia are groups of cells which all possess the potency to form the organ in question. Initially, this organ-forming potency is not equally strong at all points, but from a central maximum it shows a gradual decrease towards the periphery. Therefore, a peripheral zone, in which the potency is weak, gradually merges into material that lacks it altogether. Moreover, the area in which the potency occurs is initially much larger than the area from which the organ will eventually develop. In the beginning, therefore, the organ primordia occupy fairly large areas with vague boundaries. At this stage, these areas may overlap at the edges, the cells in the intermediate zones possessing two or more organ-forming potencies simultaneously. In the course of development, the potency becomes concentrated in the centre of the primordium. The peripheral parts, in which the potency was weaker already, now lose it altogether. Probably the "physiological competition" between the cells (p. 132) plays a role here.

A clear case of such competition has been demonstrated by Hampé (1959) in case of the tibia and fibula of the chick. If an intermediate part of the hind-limb primordium is removed, the fibula does not develop. If supernumerary material is added to this part of the primordium, the fibula becomes abnormally long. In both cases the tibia has its normal size. If the primordia of the tibia and fibula are separated from one another, the fibula becomes larger, but the tibia smaller than in the normal embryo. Apparently, in normal development the tibia has an ascendancy over the fibula; it develops first, using a certain amount of cell material, and the fibula then develops in proportion to the remaining material.

The result of all this is a clear demarcation of the organ primordia from their environment. At first, these primordia have again the character of organisation-fields. This is

proved by the fact that they possess the power of regulation, and that the organ-forming potency can be transposed
with respect to the material of the germ. In early primordia,
parts of the material of the field can be exchanged, e.g. by
inverting part of the primordium, without causing disturbances in later development. A smaller, but harmoniously
built organ will develop from an isolated part of a primordium,
so that in experiments one primordium may give rise to several
organs (cf. our discussion of the limb primordium, p. 150).
Conversely, two or more equivalent primordia can fuse so
that they produce a single harmoniously built organ of double
size. In other words, the organisation-field may be divided into
several fields, and, on the other hand, two equivalent fields may
be combined into one unity. In each organ-field, therefore, we
find, on a smaller scale, a repetition of the regularities shown
by the organisation-field of the whole embryo at an earlier
stage in development (cf. p. 128).

Initially, all cells of the field possess the same organ-forming
potency, though not all to the same degree. Later, however,
a further subdivision of the field takes place: the determination
of the parts of the organ. Under the influence of the field-
factors, chemodifferentiation takes place within the field, so
that the originally equivalent cells begin to differ in potencies.
Once this has happened, each part of the primordium can
produce one part of the organ only, e.g. only a retina, instead
of a whole eye. At this stage the parts can no longer replace
one another; disturbances and displacements of material within
the primordium are no longer regulated. The organ-forming
potency can no longer be transposed, but is tied to definite cell
groups. Following Paul Weiss (1926), we may call these
happenings *"autonomisation"* because they make the parts of
the organ autonomous so that from now on they behave as
mutually independent units. In the primary organisation-field
of the embryo, too, such a process of autonomisation takes
place. We have seen that it was this process that led to the
formation of the various organ primordia. Viewed in this light,
development may be regarded as a stepwise subdivision of the
primary field. The first step leads to autonomisation of the

organ primordia within the field, breaking it up into a number of more or less independent organ-fields. The second step is the division of the organ-fields into their parts. These parts, in their turn, behave at first as organisation-fields. Experiments, e.g. on the chick embryo, have shown that the field of the leg becomes divided into the sub-fields "upper leg", "lower leg", and "foot". On transplantation of any given part of one of these sub-fields, the graft no longer forms a whole leg, as in the previous stage, but only the appropriate part of the limb (Murray, 1926). It may be assumed that at a still later stage, these sub-fields will again be subdivided by "autonomisation" so that, e.g., the field of the foot will break up into tarsus, mid-foot and toes.

The mutual contact inductions of the parts of the germ, and the autonomisation of the parts within the organ-fields, result in a further increase of the spatial multiplicity of the embryo. More and more, the embryo is broken up into a mosaic of cell groups which differ from each other in physico-chemical constitution. This will soon express itself in the external appearance of the cells. Starting from the more or less indifferent embryonic cell type, the cells now begin to specialise for the particular functions which each of them will have to fulfil in the whole of the organism. The period of organ-formation, therefore, is followed immediately by a phase in which the differentiation of the tissues, or *histogenesis*, dominates the picture. During this period are formed the tissues of which the embryo consists, such as muscle tissue, connective tissue, nervous tissue, cartilage, etc. It is a direct consequence of the preceding chemodifferentiation; it is the visible expression of the diversity of the cells which, in an invisible form, was already present.

Histogenesis is the last important step in the realisation of the structural plan. The egg has now given rise to an organised whole, a complex of histologically different organs and tissues with fixed topographical relations. In a word, an embryo has been formed.

CHAPTER X

The later stages of development

In the previous chapter we have seen how, in the course of development, the embryo is subdivided into ever smaller groups of cells, which are more or less independent, and have different potencies. In the earlier phases of development, the whole embryo behaves as one unit. It reacts to disturbances with regulation, so that a more or less harmoniously built embryo is formed in spite of the interference. In the later stages, however, the situation is different. The power of regulation, it is true, still exists within the individual organisation-fields, but major disturbances, such as the removal or displacement of complete organ primordia, can no longer be regulated, and result in aberrant development. At this stage, the embryo can be said to consist of a mosaic of building stones which cannot be removed or displaced without damage to its development. Each of these building-stones, the organ primordia, develops more or less autonomously, and it can no longer be replaced by any of the others. This phase may therefore be called the *"mosaic stage"* of development.

That, indeed, the organ primordia are irreplaceable at this stage, is clearly proved by experiments in which the whole tail-bud of young amphibian embryos was cut off. Such embryos developed into tail-less larvae, nor was the missing part of the body replaced at later stages (Schaxel, 1922), (Fig. 52). This is the more remarkable because complete regulation would have taken place if the material concerned had been removed at a slightly earlier stage. Other material would then have formed a tail. And if, on the other hand, the tail is amputated at later stages, regeneration will occur, and a new tail will be formed by a proliferation of cells (cf. Chapter XI). The mosaic stage

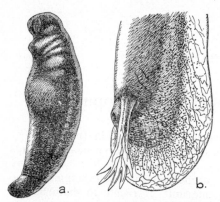

Fig. 52. Extirpation of the tail primordium of **Triton** at an early stage (*a*) results in an entirely tail-less larva (*b*). After Schaxel.

therefore, is a developmental phase of limited duration during which regulation exists no longer, and regeneration not yet.

Results similar to those described for the tail primordium have been obtained with other organs as well at this stage. Complete extirpation of a limb primordium, for instance, results in lasting absence of the limb concerned, both in amphibians, and in the chick embryo (Fig. 53). The same applies to eyes, gills, etc., and even to the blood of amphibians: removal of the rudiment of the blood cells leads to the development of larvae without blood corpuscles (Frederici, 1926).

The high degree of independence of the organ primordia is also demonstrated by their behaviour on transplantation. In the previous chapter we have seen many examples of the fact that, once they have been determined, organ primordia grafted into a foreign environment can continue their development in complete independence, and that they can in this way reach a high degree of differentiation. This, moreover, is true not only of amphibians, but of other animals at the same stage as well. Organ primordia of the chick embryo, grafted into the egg membranes of another, much older embryo, will often continue their development, and differentiate into the organ they were destined to form. The same applies to fishes. Here organ primordia have been grafted onto the yolk sac of other embryos. Organ rudiments of mammals, e.g. rabbits and rats, can be grafted on to the omentum in the abdominal cavity of a rabbit. In that case, and, what is more, even after transplantation

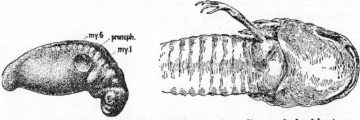

Fig. 53. Extirpation of the fore-limb primordium of **Amblystoma** at an early stage (left), leads to the production of a larva (right), in which the limb concerned is completely absent. After Harrison.

on to the egg membranes of a chick embryo, they, too, were found to continue their differentiation.

The case of the mesonephros primordium of the chick embryo deserves special mention. In normal development, the mesonephros of the chick becomes reduced from the tenth day of incubation onwards, its function being taken over by the metanephros. If grafted on to the egg membranes of an older embryo, the primordium first differentiates into a typical mesonephros. Reduction, however, sets in at about the same time at which it would have happened in the normal embryo (Danchakoff, 1924). This shows that not only the nature of the progressive differentiation, but its duration as well is determined in the organ primordium. Holtfreter (1931) found a similar case in amphibians. Here, explanted endoderm of the prospective mid-gut remains compact for several days, but it becomes broken up into loose cells at the time at which, in the normal embryo, the yolk-rich endoderm cells would have been expelled into the mid-gut lumen.

Apart from transplantations, explantations have also been applied to the study of the properties of organ primordia. This method proved that in many cases the primordia are capable of independent differentiation outside the body, provided they are cultured in a suitable medium. The heart rudiment of amphibians, for instance, developed *in vitro* into a pulsating heart in which the various parts of a normal heart were clearly distinguishable (Stöhr, 1924). Explanted parts of cephalopod (cuttle-fish) embryos also proved able to continue their differentiation in the normal way (Ranzi, 1931).

A somewhat different situation obtains in tissue cultures. Here the tissues are kept in a medium specially provided with food and growth promoting substances. This stimulation of growth inhibits differentiation. Therefore, embryonic tissues are usually unable to differentiate in these cultures, and such structure as may already be present in the tissues will even be reduced so that the cells become less differentiated in their external appearance. However, under certain circumstances, differentiation is possible in tissue cultures. Gaillard (1931), for instance, has shown that the differentiation of bone-forming tissue is promoted by the successive addition to the culture of a series of extracts from embryos of steadily increasing age.

In this context, it is important to note that the loss of visible cell differentiation in tissue cultures does not imply that the cells really relapse into an undifferentiated, embryonic condition. It is true that cells determined into different directions e.g. bone-forming cells, and cells of the heart-primordium, in tissue cultures become indistinguishable in external appearance, but even after a very prolonged stay, they still retain their different properties, and as soon as circumstances permit they return to their particular course of differentiation. There is no real "*de-differentiation*", therefore, but only "*modulation*" (P. Weiss, 1950).

We have already seen that the beginning of the mosaic stage is soon followed by visible tissue differentiation. After this, *growth* becomes gradually more important. Admittedly the embryo may have increased in size before this stage, but this was mainly due to water uptake, and was more a swelling process than real growth. True growth cannot begin until differentiation has proceeded so far that the embryo is able to take up food independently, or at least, until its blood circulation is well enough developed for the transport of food from the yolk, or from the maternal tissues, toward the embryo to be possible. In other words, the period of differentiation generally precedes growth.

Though growth is highly dependent upon such external

factors as temperature and food supply, yet each species has its own rate of growth. This was clearly shown by experiments in which the primordia of a limb were exchanged between embryos of two newt species, viz. *Amblystoma punctatum* and *A. tigrinum*, which at their later stages show different growth rates. It was found that the grafted limbs retained their own growth rates, and did not adapt themselves to the slower or faster rate of the host (Harrison, 1924), (Fig. 54).

Moreover, the various organs also have their own characteristic growth rates. This causes changes in the body proportions in the course of development.

Finally, various organs influence the growth of other organs. This was demonstrated by combining, by means of a heteroplastic transplantation, a lens of *Amblystoma punctatum* with an eye-cup of *A. tigrinum*. The more rapidly growing *tigrinum* eye-cup stimulated the more slowly developing *punctatum*-lens to faster growth.

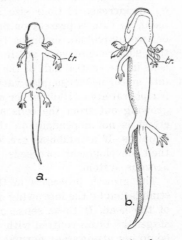

Fig. 54. Larvae of (*a*): **Amblystoma punctatum** and (*b*): **A. tigrinum**, between which the left fore-limb primordia have been exchanged at an early stage. After 50 days, the transplanted limbs are about the same size as those of the animal from which they originate. After Huxley and de Beer.

The same is true of a *tigrinum* lens when combined with a *punctatum* eye-cup. A mutual adaptation in size and growth rate occurs between the two parts (Harrison, 1929; Twitty and Schwind, 1931). In Rotmann's experiments, too, the originally maladjusted eye-cup and lens of the heteroplastic combinations (see above, p. 141) were later on found to show mutual adaptation in size by differential growth.

This mutual influence is also clearly expressed in the central nervous system. The size of the spinal ganglia depends to a great extent upon the size of the peripheral area with which they are connected. Increase of this area, e.g. by the implantation of a supernumerary limb in the flank, causes an increase in the number of sensory neurones in these ganglia, and thereby an increase in their size (Detwiler, 1920). If, on the other hand, no skin is present on one side of the body, as in the case of two amphibian embryos fused side by side, the ganglia on this side remain very small (Detwiler, 1926). The number of cells in the ventral half of the spinal cord, that will differentiate into motor neurones, also varies with the size of the innervated peripheral area (Hamburger and Keefe, 1944). The nerve fibres growing out from the cells seem to play a role here, but the effect is not dependent on the normal conduction of nervous impulses. If amphibians are kept under permanent anesthesia, their development proceeds normally, including that of the nervous system.

The growth processes in the brain are partly governed by stimulation by the ingrowing nerves from the large sense organs of the head. If these sense organs are extirpated at an early stage, the brain centres with which they are normally connected do not attain their normal size. Conversely, the implantation of a supernumerary eye or olfactory pit will cause an overdevelopment of certain parts of the brain, if the nerve coming from the graft penetrates into the brain of the host.

The growth of peripheral nerves towards their end organs seems to be governed by attractions exerted on the nerves by strongly growing organs. A limb transplanted into the flank will establish connections with any nerve that happens to be in the neighbourhood, even though normally this nerve has nothing to do with limb innervation. The nerve fibres enter into the limb and become connected with its muscles, so that in many cases the latter will function more or less normally (Detwiler, Weiss). An eye or an olfactory pit, transplanted into the flank, also attracts the neighbouring spinal nerves (Detwiler, 1927), though in this case no functional connections between nerve and organ are formed.

Once the connections of the nervous system with the various organs have become established, the former begins to influence the development of the latter. It is true that practically normal differentiation can take place in amphibian limbs without nerves, but such limbs remain smaller than normal ones (Hamburger, 1929). Evidently, the nervous system has a trophic influence which intensifies growth. In other cases, however, the connection with the nervous system is indispensable for the maintenance of the structure of the organ. Marked atrophy occurs, for instance, in muscles as soon as the motor nerves are severed. This is even more true in the case of the lateral line sense organs and the skin taste buds of fishes, which degenerate entirely after transection of the afferent nerve (Olmsted, 1920). This influence of the nervous system is probably due to the secretion of a substance by the nerves.

The development of the blood circulation also has important consequences for further development, as it supplies the various organs with the food substances needed for their growth. The distribution of these substances between the organs will depend upon the total quantity that is available. The body proportions of maximally fed animals may differ sharply from those of starved specimens. In extreme cases, certain organs or parts of the body may be completely reduced in consequence of want of food.

The possibility of hormone transport is another important consequence of the development of the vascular system. Hormones are substances produced by certain glands, which become distributed throughout the body by the blood so that they can exert their influences on the development of other organs in remote places. Hormones play a very important role, in particular in the later stages of development, and in the adult animal. A more extensive discussion of these phenomena lies outside the scope of this book.

Finally, in the later stages, the function itself of each organ has an important influence on its development. Once differentiation has proceeded so far that the organs begin to take over their own specific functions, new processes are started under the influence thereof. These processes influence the structure of the organs, and modify their growth and histological differentiation, or the arrangement of their cells and tissues. We shall now briefly review this *"functional stage"* of development.

A well known example of the influence of function on organ growth is the great increase in the size of striated muscle as a consequence of heavy muscular work. Another phenomenon in this context is the so called "compensatory hypertrophy" after removal of part of an organ. If one kidney of a vertebrate animal is extirpated, the other kidney increases in size.

Not all cases of such hypertrophy can, however, be explained by the excessive functional load placed upon the remaining fraction of the organ system. If contralateral appendages in worms, or testes in mammals, show excessive growth after removal of the organs of the other side, such an explanation does not hold. Therefore, P. Weiss has put forward a theory of specific growth control which may account for such phenomena. According to this theory, growth of each specific cell type is catalyzed by specific "templates" within the cells. Each cell also produces "anti-templates", which can inhibit the templates by combining with them into inactive complexes. The anti-templates are released from the cells and get into circulation. As the concentration of anti-templates in the extracellular medium increases, growth in all cells of that type will decline, and finally cease. Removal of part of an organ, by diminishing the production of anti-templates, will cause resumption of growth till a steady state is restored (Weiss, 1955).

The influence of the functional relationships on the arrangement of tissue elements is clearly illustrated by Weiss's (1929-33) experiments. He exposed tissue cultures of connective tissue cells (fibroblasts) to local tension. The cells then became arranged in accordance with the direction of the lines of tension, and multiplied more rapidly in this direction. In normal development, such reactions play a role in the origin and regeneration of tendons, etc., as was shown by Lewy's experiments as early as 1904. The detailed architecture of bones obeys the same law. The trabeculae (bone bars) in spongy bone are arranged in the direction of the stresses normally acting on the bone. If the direction of these forces is modified, e.g. by the extraction of a molar from the jaw, the direction of the trabeculae is changed accordingly. The growth of bones also depends upon their function. A motionless limb which is not subjected to any stress is retarded in growth, as compared with a normally functioning limb.

Function is of paramount importance also for the final structure of the vascular system. With the exception of the very first blood vessels, which differentiate locally, all others arise in the course of embryonic development by the branching of existing vessels. Initially, massive "buds" of endothelium cells are formed. These extend into the tissues and soon become hollow. Manifold connections between the outgrowing blood vessels are then formed, so that a network develops. At first, all these vessels are of about the same calibre. The same process takes place in regeneration. This primitive network differentiates into the definitive vascular system in the following way. Some of its meshes disappear, others become capillaries, but the lumen of a few channels becomes widened, and their walls thickened so that they become the major arteries and veins. Clark and collaborators have studied this process in living animals, viz. in the transparent caudal fins of frog larvae, and in specially constructed transparent chambers, let into the ear of a rabbit. They found that the fate of each part of the original network depends on the currents which prevail there because of the pressure differences in the system. Where the blood stagnates, the diameter of the vessels decreases, and finally they disappear. Strong perfusion, on the other hand, causes a widening and thickening of the vessels concerned, leading to their differentiation into major blood vessels. In this way the blood stream itself models the eventual pattern of the blood vessels.

The gills of larval newts may be taken as a last example. In water with a low oxygen content, these organs are large and strongly branched; the overlying skin is thin, and they are well vascularised. In water with a high oxygen content, on the other hand, where skin respiration can easily take place over the whole body surface, the gills are poorly developed. Here, again, the structure becomes adapted to the functional requirements under the influence of the function itself.

CHAPTER XI

Regeneration

Regeneration is the term applied to a process of restitution which occurs after removal of a part of the body and which results in complete or partial replacement of the lost part.

If an animal of simple structure, e.g. a planarian, is cut into two transversely, a bud-like outgrowth will within a few days develop on the cut surface in each half. This is the so-called *regeneration-bud*. It grows steadily, and replaces the lost part of the body, i.e. a new fore-part is formed on the caudal half, and a new hind-part on the anterior half (Fig. 55). The position of the cut is of no, or at most of secondary importance. Whether it is made in the middle, or in the front or hind part of the worm, the result is always the same. This shows already that the quality of the regenerate is not determined exclusively by the nature of the regenerating material. The situation is not

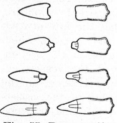

Fig. 55. Regeneration of rostral and caudal halves in the flat worm **Polycelis nigra**, after 0, 7, 9, and 14 days respectively. After Dresden.

such that, e.g., in the front half of the body "tail-forming" cells or substances are present, and "head-forming" material in the posterior half. At each level in the body, both a head and a tail can be produced. What will in a given case develop from the regeneration-bud, depends only on whether the cut at which this bud is formed faces forwards or backwards. In other words, the original polarity of the worm determines the result of the regeneration. Evidently, this polarity has remained intact in both pieces formed by the transection of the worm. This is true even if the worm is divided into several parts by a series of transverse cuts. The original polarity is retained by each of these parts, and each of them will regenerate a head at its front edge, and a tail at its hind edge. Only in exceptional cases can this inherent polarity of a fragment be suppressed or reversed by certain treatments. In some cases, however, the situation is more complicated. In earthworms, such as *Eisenia foetida*, a number of regions with different powers of regeneration can be distinguished (Gates, 1950). In the foremost six segments of the body, a head is regenerated at a front edge, but no regeneration takes place at a hind edge. In the succeeding zone of about 11 segments, heads are formed by both front and hind edges. From segment 21 to segment 34, a head or a tail may be formed at either edge. Next comes a region, extending to segment 54, in which only tails are regenerated at both edges. Finally, in the zone behind segment 54, a tail is formed by a hind edge, but no regeneration occurs at a front edge.

It is important to study the nature of the regeneration process somewhat more closely. We have already mentioned that the formation of a regeneration-bud is the first visible consequence of the transection. This consists of cells of a very indifferent character. The origin of this material is not in all cases sufficiently well known. In some lower organisms, un-differentiated cells very rich in ribonucleic acid are present at various places in the body. After transection, these migrate into the wound area, and accumulate there. Miss Dubois (1949), for instance, has demonstrated this in the case of *Planaria*. Regeneration can be suppressed here by ir-radiation with X-rays; the irradiated animals succumb after a few weeks. But if only the foremost two fifths part of the body is irradiated, and the head is then amputated, regeneration does occur, though it begins a month later than in non-irradiated animals. Miss Dubois has been able to prove that this is due to the fact that the regeneration cells, or *neoblasts*, migrate from the posterior, non-irradiated part of the body into the wound area. In the course of this migration, they must traverse the irradiated zone. The wider this zone is, the more regeneration is retarded. This migration is not spontaneous; it takes place only after a wound has been made, and it is always directed towards the wound. The cells can move both forwards and backwards, and sideways. The stimulus that causes the onset of migration spreads throughout the body; it disappears four to five days after the wound has been made. Experiments in which a regeneration-bud was explanted *in vitro* together with a piece of minced adult planarian tissue seem to show that this stimulus is humoral in nature (Wolff, 1961). Once started, the migration continues until a regeneration blastema has formed in the wound area. Apart from the capacity to form a regeneration-bud, the neoblasts possess the power to repair other damage in the body. They are omnipotent, i.e. they can form any other tissue.

Migrating neoblasts have also been found in some annelids, but not in all. Whether the so-called interstitial cells of hydroid polyps play a similar role in the restitution processes occurring in this group (cf. below), is still a matter of controversy.

It is probable that in other cases the regeneration cells are formed by the de-differentiation of tissue cells which thereby acquire a more embryonic character, and an increased power of division.

Once the regeneration-bud has reached a certain size, an outwardly visible differentiation begins to take place in this material, and it becomes possible to see what will develop from it.

Now it has been found that in many annelid worms the regeneration-bud which develops at a front edge will always differentiate into a fixed number of segments, which agree with those found in the rostral end of the body, the "head" of a normal worm. The number of segments in the regenerate is entirely independent of the place of the cut. In the polychaete worm *Sabella*, for instance, three segments are always formed, the foremost of which carries a crown of branchial palps, even if the regenerating fragment belongs to the caudal end of a normal worm (Fig. 58). This shows that here regeneration does not lead to restitution of the missing part of the body, but produces only a new apical end by an autonomous differentiation. In contrast, the type of differentiation in regeneration buds at a caudal edge depends upon the nature of the tissues of the fragment, and here the lost part of the body is completely replaced.

The same applies to planarians. Here again, regeneration at

a front edge, irrespective of its position, produces only a head, whereas that at a hind edge leads to complete restitution of the missing parts.

Similar phenomena are found in the restitution processes that occur after the transection of hydroid polyps, though here as a rule we find no true regeneration, but rather a "reorganisation" of the old tissues, which are subjected to a morphological dedifferentiation, followed by new differentiation. If a stem fragment of such a polyp is isolated, it will change into a new hydranth provided the size of the fragment is sufficient. Smaller parts produce only the apical parts of a polyp, but these apical parts are of normal size.[1] Evidently, the process of reorganisation begins with the formation of the apical parts, and proceeds from there in the direction of the base, until all the available material has been used (Fig. 56).

The size of the apical parts formed depends more or less upon the external circumstances. In *Planaria*, for instance, only a small head is produced by a regeneration-bud at low temperatures, or under the influence of anesthetics. High

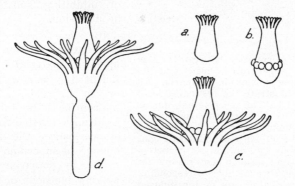

Fig. 56. Reorganisation in the hydroid polyp **Tubularia**. A very small part produces only (*a*): the apical part of a polyp; if more material is present, the more basal parts are formed accordingly (*b-d*).
After Child.

---

[1] In many cases, apical parts arise at both ends of the stem fragments. We shall not enter into this complication.

temperatures, on the other hand, result in the production of very big heads (Child, 1915), (Fig. 57). In the hydroid *Tubularia*, the size of the hydranths produced depends on the oxygen tension. Here, too, narcotics cause a reduction in size, which however can be compensated by increased oxygen tension (Barth, 1944). In normal regeneration, the role played here by anesthetics is probably duplicated by inhibiting substances formed in the tissues. After transection, regeneration is started by the better oxygen supply to the wound area, and by the disappearance of inhibiting substances by diffusion.

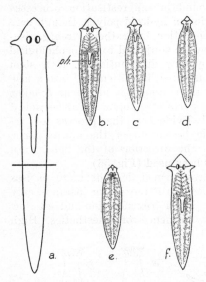

Fig. 57. Correlation between the size of the regenerated head and its organising power in **Planaria**. All figures illustrate the regeneration of a caudal half of the original, cut as in (*a*); (*b*) regeneration under normal circumstances; (*c-e*) regeneration under the influence of increasing concentrations of narcotics; reduction in size and development of the regenerated head, and corresponding reduction in the size of the induced pharynx (*ph*), and in the distance between head and pharynx; (*f*) regeneration at optimal temperature larger head and pharynx, and greater distance between pharynx and head. After Child.

In some cases, local stimulation without removal of material is sufficient for a new head, or, in hydroids, a new hydranth, to be formed. Goldsmith (1940), for instance, caused the development of supernumerary heads in planarians by making incisions or local burns in the rostral part of the body. In the earthworm a supernumerary head may develop laterally in the front part of the body, if a wound is made there, and a cut end of the ventral nerve cord is brought into the wound (Avel, 1947). Evidently the nerve cord has a stimulating or activating influence, for the wound itself does not cause head formation at this place.

It may be assumed that in the differentiation of organs from a regeneration blastema similar mutual influences to those we have encountered in the development of embryos play a role. In fact, the origin of eyes in a regenerate in the flat-worm *Polycelis* proved to depend on a substance produced by the cerebral ganglion, which diffuses through the body. Its concentration decreases in a posterior direction; moreover, it seems to be taken up in an inactive form within the cells of the posterior part of the body (Lender, 1956). Only the cells of the region normally bearing eyes have the competence to react with eye-formation to the presence of this inducing substance.

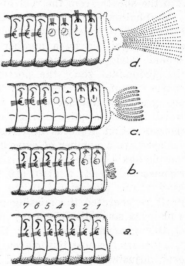

The head formed by regeneration at the front edge of a caudal fragment of the body does not at first harmonise with the old tissues of the fragment, because certain intermediate parts of the body and their organs are absent. The normal structure of the organism is restored by a second process, subsequent to that of regeneration in the proper sense. This process is called *morphallaxis*. In its course, regenerate and fragment mutually

Fig. 58. Four stages (*a, b, c, d*) in the regeneration of the rostral end in **Sabella**. As regeneration proceeds, the foremost 4 segments change from the abdominal into the thoracic type, by the loss of their old hooks and bristles and the formation of new ones in another position. This change proceeds in a rostro-caudal succession.
After Berrill and Mees.

adjust their sizes (Fig. 55), and, moreover, the newly formed head influences the adjacent tissues in such a way that the missing parts are formed here by "reorganisation" of the tissues. If, for example, the caudal part of a planarian has regenerated a new head, the pharynx which in planarians occupies the middle of the body, at first is absent. By subsequent morphallaxis, however, a cavity is formed in the old tissue of the original fragment. This develops into a new pharynx sheath, and breaks through to the exterior, forming a new mouth. The pharynx itself is formed by the regeneration blastema and grows into the sheath from the rostral side (Van Asperen, 1946).

Morphallaxis is very clear in *Sabella*. In this worm, the foremost 5 to 11 segments, the so-called "thoracic" segments, are distinguished by a number of characters from the more caudal "abdominal" segments. If the animal is transected in the abdominal zone, the posterior part regenerates a head consisting of three segments. This, however, is followed by changes in a number of the foremost abdominal segments, which thereby acquire thoracic characters (Berrill, 1931). This reorganisation proceeds in a rostro-caudal direction (Fig. 58).

There are a number of experiments that leave no doubt that morphallaxis is due to "organising" activities of the previously regenerated head. Santos (1929) transplanted parts of the head of one planarian into the caudal region of another. The graft regenerated the missing parts of the head, and induced a pharynx some distance away in the host tissues. The direction of this pharynx agreed with that of the implanted head. In many cases the host tissue even supplied a complete new caudal part, adjusted to the implanted head (Fig. 59). Similar effects were obtained in the case of heteroplastic transplantation between two species of *Planaria*.

These experiments were repeated and extended by Sengel (1953). After a pharynx had been induced in the host tissue, the grafted head was excised. After some time, the induced pharynx was removed too. This was now regenerated in its original direction, even when the latter was reversed with respect to the polarity of the host. It is concluded that the graft did not merely induce a pharynx, but has transformed

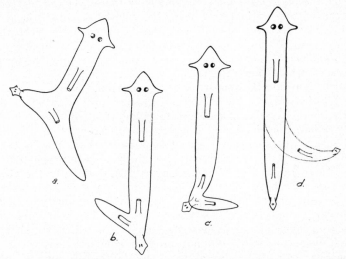

Fig. 59. Organising activity of a transplanted head in **Planaria**. (a) a laterally implanted graft has induced a lateral outgrowth and a secondary pharynx; (b-c) a subterminal graft has induced an outgrowth (directed forwards), and two pharynges; (d) a terminal graft has caused a reversal of polarity in the caudal part of the host, and a secondary pharynx; in this last case the hind part was capable of autonomous movement (dotted outline), i.e. it behaved as an independent individual. After Santos.

the adjacent region of the host into a pharyngeal zone, which has thereby got the properties of a normal pharyngeal region.

In hydroids, too, the existence of such organising activities has been demonstrated. Here they are exerted by the apical parts of the polyp. In *Hydra,* the peristome (i.e. the area around the mouth) of one polyp, implanted laterally into the body of another, induces the formation of tentacles and the production of a bud.

The comparison is obvious between these organising influences of regenerated heads or apical parts and the activity of organisers in embryonic development. Here, again, the situation can be described as an "organisation-field", originating from the regenerate and spreading from there over the adjacent areas of the original fragment; the role of organiser is here played by the regenerate. Under the influence of the field, the cell material undergoes certain changes characteristic of morphallaxis, changes which adapt regenerate and fragment to each other. Therefore, the study of regeneration may deepen our insight into the properties of organisation-fields.

First of all, the field in certain cases appears to adapt its size to the available material. In a fragment of the ascidian *Clavelina,* for instance, complete de-differentiation of the tissues into a compact mass of cells may occur. This is then followed by fresh differentiation into a well-proportioned *Clavelina* of smaller size (Huxley, 1926). Evidently the field in this case harmoniously imposes itself on such material as is present; its size is adjusted to the available quantity of cells. However, that this is not always so, may be seen from the case of hydroid polyps described above, in which fragments that were too small did not produce complete polyps on a smaller scale, but only parts of hydranths of normal size. On this point, therefore, the field may behave in either of two different ways. As yet it is entirely unknown which factors are responsible for the choice between the two.

Further, it was found that the area eventually occupied by the field depends upon the "state of activity" of the regenerate acting as organiser. This, in turn, can be influenced by external factors. If, under the influence of cold or narcotics, the re-generated head of a planarian has remained small, the pharynx which arises by morphallaxis under the influence of the field also remains smaller, and it is formed at a shorter distance from the rostral end. There is a direct correlation between the size of the head, that of the pharynx, and the distance between the two (Child, 1915), (Fig. 57). Similar phenomena have been found in hydroids. In *Sabella,* illumination of the regenerating head causes a strong expansion of the field. In darkness, only about four abdominal segments develop into thoracic segments during morphallaxis, but this number may be increased to as many as eighty by illumination (Berrill and Mees, 1936).

The organising power proved not to be a privilege of the

extreme rostral, or apical, part of the body. As a rule, each part of the body has an organising influence on the more caudal parts. A fragment from the middle of the body of a planarian can regenerate a tail at its hind edge, even if no regeneration of a head takes place at its rostral end. Even a pharynx may be formed in this case, at least if the fragment originated from the pre-pharyngeal part of the body, but not if it originated from the post-pharyngeal region. This agrees with the observation by Okada and Sugino (1934) that not only transplanted heads, but also more caudal parts of the body of a planarian have inductive effects when grafted into still more posterior regions of another individual.

The same applies to the polyp *Corymorpha*. Here parts of the stem, grafted laterally into the stem of another polyp, in a number of cases induce a new hydranth. This happens more often, according as the original position of the graft in the stem was more apical (Child, 1929), (Fig. 60).

The inductive power, therefore, is not a property of one definite, specific tissue only, but it is due to a physiological condition the intensity of which decreases as a gradient from the rostral, or apical, towards the caudal, or basal, end of the body. In other words: each part of the field influences the more caudal (basal) parts, but is itself influenced by the more rostral (apical) parts.

Little is known so far about the physical nature of this gradient, though it is obvious to think of electro-physiological phenomena in this connection. The work of

Fig. 60. Organisation by transplanted parts of the stem in the hydroid polyp **Corymorpha**. (*a*) an apical part of the stem induced a complete new hydranth in 48 hours; (*b-c*) more basal parts of the stem induced smaller (*b*) or abnormal outgrowths (*c*) in the same time. After Child.

Moment (1946-49) is highly important in this respect. We have seen above (p. 170) that a regeneration-bud at a hind edge as a rule completely replaces the lost part of the body. Now Moment has shown that during caudal regeneration in the earthworm *Eisenia foetida* the formation of new segments stops when the total number of segments has reached an approximately normal value. In a normal worm, there is an electric potential difference between the rostral and caudal ends of the body. This potential is destroyed by amputation, but during regeneration it is gradually built up again until it has regained its original value, which happens at the very same moment at which the formation of new segments comes to a stop. On this basis, Moment concluded that the production of new segments is brought to a close by an inhibitory action exerted by the electric field as soon as this has attained a certain degree of intensity.

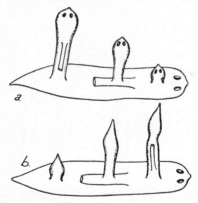

Fig. 61. The effect of discontinuities in the gradient in **Planaria.** (*a*) the growth caused by grafted heads is the more marked when the head is implanted more caudally; (*b*) on the other hand, transplanted hind ends produce strongest growth in the rostral parts of the host. After Schewtschenko.

The gradient character of the regeneration field often also finds expression in the rate of regeneration. In posterior regeneration of *Eisenia* this rate decreases as the level of transection becomes more posterior (Moment, 1953). In planarians, the time required for the regeneration of a head increases both anteroposteriorly and mediolaterally, so that there is a "time-graded regeneration field", which has a characteristic shape for every species (Brøndsted, 1955).

Two parts originally situated at different levels in the gradient-field can be brought into contact by transplantation. In this case, a strong proliferation of tissue occurs at the boundary (Okada and Sugino). The greater the discontinuity in the gradient, the more marked is this proliferation. This was beautifully demonstrated in an experiment by Schewtschenko (1937), who grafted heads of planarians into different regions of the hosts. Implanted caudally, the heads caused a very strong outgrowth, but in grafts in more rostral areas the effect became gradually weaker. Transplantation of caudal parts had exactly the opposite effect: strong proliferation in rostral areas, much weaker proliferation more caudally (Fig. 61).

One of the effects of the "dominant" area (Child) on the subordinate parts of the field is of an inhibitory nature. Miller (1938) transplanted the head of a planarian into the trunk area, and then transected the host just rostral to the graft. No head was then regenerated at the front edge, because the graft repressed such regeneration. A head grafted into the trunk sometimes moves in a forward direction because (1) strong proliferation takes place at the caudal side of the graft, and (2) the tissues in front of the graft are resorbed. This movement goes on until the head has found its way to the rostral end.

It has been proved that the inhibitory action of a dominant region may be due to the production of inhibiting substances. Rose, Steinberg and others have shown that in hydroids the hydranths produce substances which diffuse through the body cavity and suppress the formation of other hydranths. Tardent showed that these inhibitory capacities of a hydranth are acquired during the process of regeneration, and increase in the course of development. On the other hand, susceptibility of the regenerates with respect to the inhibitory action decreases with differentiation. Therefore, older regenerates inhibit younger ones, but self-inhibition of a regenerate is prevented (Tardent, 1960). In planarians, the regeneration of the cerebral ganglion is suppressed by substances present in minced head tissue.

The apical "organiser", therefore, has a twofold influence. It prevents the formation of new apical parts, and it forces the other cells within the field to form subordinate organs in accord with their place in the field.

If a tissue is to form a new apical region, it is necessary for it to be removed from the inhibitory influence of the existing apex. This can be done by cutting the latter away, but in normal development it may occur as a consequence of growth. The field emanating from the "dominant" apex has a limited extent, and consequently part of the tissue can be removed from its sphere of influence by growth — so-called *"physiological isolation"* (Child). This tissue will then form a new apical region. This is the law governing the occurrence of new buds in colony-building organisms, such as hydroid polyps. We have seen that the extent of the field depends upon the state of physiological activity of the tissues. The distance between buds formed under favourable conditions of food, temperature, etc., will therefore be greater than that between buds formed under less favourable conditions. This may completely modify the appearance of the colony as a whole (Child, 1929). A similar phenomenon occurs in some planarians which reproduce asexually by transverse division. Here, at a certain distance from the original head of the individual, the caudal part forms a new head. If, under unfavourable circumstances, the size of the old head is reduced (see p. 171), the field emanating from it extends less far, and division takes place at a much smaller size of the body.

Child has demonstrated in many experiments that the dominant apical parts are often highly sensitive to noxious influences, and that this sensitivity decreases in a caudal (or basal, respectively) direction. He ascribed this to differences in the "physiological (i.e. metabolic) activity" of the tissues, which would be highest in the apical parts, and decrease gradually from there. Similar results were obtained with another method for the demonstration of differences in the intensity of metabolism, viz. measurements of the velocity of decolorisation of certain redox dyes, such as methylene blue, at low oxygen pressures. From these differences in physiological activity, Child explained such morphogenetic influences as "dominance" and "organisation", exerted by the apical part.

It is doubtful, however, whether in doing so he did not confuse cause and effect. In other words, the morphogenetic dominance of the apical region might be the primary phenomenon, and the difference in physiological activity of the tissues its consequence. So far it has not been possible to decide which of the two hypotheses is correct.

On the basis of experiments on regeneration in hydroid polyps, Spiegelman (1945) has developed a theory which reduces the dominance of a cell group over its neighbours to "physiological competition" between the cells by the absorption of food substances and the secretion of noxious waste products. It is possible that several phenomena in this field can be explained in this way. However, in an investigation on the mutual influences of rostral and caudal regenerate in transverse sections of *Euplanaria lugubris,* Raven and Mighorst (1948) showed that in this case the facts are not what one would expect on the basis of the theory of "physiological competition".

Brøndsted (1955) has put forward another theory to account for these phenomena. It starts from the notion of a "time-graded regeneration field" mentioned above (p. 178). In such a field every cut will expose a surface in which some place has the highest regeneration rate. Here regeneration of the head starts. At the same time it inhibits neighbouring parts from exercising their ability to make heads themselves. A similar theory of specific inhibition has been further elaborated by Rose (1957), who applies it to differentiation in general. According to this view, pattern arises as the products of differentiating regions act on other areas and suppress like differentiation there. The dominated areas, prevented from developing as they would have done in isolation, are reduced to a lower order of differentiation. These in turn suppress like differentiation in other cells, and so on. By such a series of specific inhibitions, differences may arise in a system of originally equipotential cells.

All examples discussed so far referred to so-called "total regeneration", provoked by transection of the whole body. However, the study of the regeneration phenomena which occur after the removal of subordinate parts of the body has further contributed to our insight into the mode of action of organisation fields. Interesting results were obtained in particular in experiments on the regeneration of limbs and tail in amphibians.

If a limb or the tail of a larval or adult newt is cut off, a regeneration bud, consisting of more or less indifferent cells, is formed in the wound area. The origin of this material and the processes resulting in the formation of the regeneration blastema have in recent years been the subject of many investigations. It has been found that, immediately after the amputation of a limb, a marked de-differentiation takes place in the tissues of the stump near the wound area. A partial disintegration of skeleton and musculature occurs, and this process supplies great numbers of free cells of a more or less indifferent appearance. In the course of the next few days, a regeneration blastema forms in the cut.

It is the prevailing view that the cells of this blastema do not as, e.g., in planarians originate from a store, somewhere in the body, of cells which have retained embryonic properties, and which migrate towards the wound. The cells of the blastema are supposed to arise locally by de-differentiation of the tissues near the wound. The following type of experiment supports this view. The distal part of a limb is replaced by a limb graft from another individual, the cells of which can be distinguished from those of the host, e.g. by their pigmentation, or haploidy. Thereupon, the graft is amputated almost completely, but so that the cut goes through the grafted tissue only. In this case, the cells of the regenerate have in every respect the character of the grafted tissue, even though the normal tissue of the host begins at a short distance from the wound. Furthermore, it has been demonstrated that the formation of the blastema begins even before there is any considerable increase in the number of cell divisions in the neighbourhood of the wound, so that it is clearly not formed by multiplication of cells.

On the other hand, recent experiments have shown that a certain amount of migration of regeneration cells may also occur in amphibians. Like in lower animals, the power of regeneration of amphibian limbs may be completely suppressed by irradiation with X-rays. If the distal part of a limb is irradiated, and then amputated through the irradiated region, at a distance of from 1 to 5 mm from the non-irradiated base, then a delayed formation of a regeneration bud may occur, probably by cells which have migrated from the healthy non-irradiated tissue through the irradiated part of the stump. The regeneration cells probably derive from fibroblasts of the connective tissue (Wolff, 1961).

Once a regeneration blastema has been formed at the wound, this seems to act on the stump, bringing the process of de-differentiation of the old tissues to a stop. Butler concluded this from experiments in which regeneration was studied after irradiation of the limb with X-rays. After strong irradiation, the power of regeneration had disappeared completely. Amputation of the limb led to marked de-differentiation of the stump tissues, but no blastema was formed, and de-differentiation continued until the whole stump had been resorbed. The same thing happened after the amputation of a limb which had previously been denervated by transection of its nerves (Butler and Schotté, 1941). Apparently the presence of nerves is indispensable for blastema-formation. Singer (1942-49) has shown that both motor nerves and sensory nerves play a role here. A certain minimum number of nerve fibres per unit area of the wound is required for the normal progress of regeneration. It appears that a considerable invasion of nerve fibres into the epidermis of the regenerate takes place in the early stages of regeneration. A richly innervated epidermal cap of cells develops over the wound surface. Blastema cells accumulate beneath the epidermal cap. In the absence of nerves no apical cap formation takes place, and the blastema cells fail to accumulate (Thornton, 1956) (Plate XVI[1]). Experiments by Butler and Schotté (1949) have proved that the presence of nerves is necessary for the first phases of regeneration, during which the blastema is formed and determined, but that the later morphogenesis, differentiation, and growth of the regenerate are independent of the nervous system.

---

[1] Facing page 155.

In tail regeneration a similar influence of the nervous system seems to play a part: both in lizards and in newts regeneration of an amputated tail only occurs, if the spinal cord is present in the region of the cut.

The de-differentiation which supplies the material for the regeneration blastema can be inhibited not only by the formation of this blastema, but also by a premature growth of skin over the wound. This may be the explanation of the difference in regenerative capacity between urodeles and anurans. In urodeles, an amputated limb is completely regenerated, both in larvae and in adults, whereas in most anurans the capacity to regenerate is permanently lost during metamorphosis. Now it has been found in adult anurans that the skin soon overgrows the amputation wound, and that the wound is covered by a scar tissue consisting of coarse connective tissue fibers. No regeneration blastema is then formed. It is of great practical importance that de-differentiation has been successfully provoked by mechanical (Polezajev, 1939-41) or chemical stimulation (Rose, 1942-45; Polezajev, 1945-46). In adult anurans a certain degree of regeneration has been produced in this way. The presence of nerves is a necessary prerequisite also in this case. It has even been possible to induce regeneration of the forelimb in the postmetamorphic frog by augmenting the normal nerve supply of the stump in experiments, in which nerves of the hindlimb were deflected to the amputation stump of the forelimb (Singer, 1954). The loss of regenerative capacity in adult anurans might therefore possibly be explained by changes in the nervous supply of the limb tissues, or by a decreased responsiveness of these tissues to the nerve influence.

Man and the mammals do not possess the capacity for spontaneous regeneration, but these investigations have revealed methods of stimulating the regeneration of lost parts of the body which might bring regeneration in this group within the realm of practical possibilities.

We shall now discuss some details of the later development of these regeneration-buds. Here all the missing parts are replaced by the regenerate, in contrast to what we have seen earlier in the case of total regeneration at a front edge. The differentiation in the regenerate adapts itself completely to the remaining organs of the stump. Moreover, we find no trace of the influence of the regenerate on the stump which was so characteristic of total regeneration. In the present case, regeneration is not followed by morphallaxis. Whereas, in total regeneration, the main interest was focused on the influence of the regenerate on the fragment, here, in contrast, we must study the influence of the stump on the regenerate.

First of all, it appears that the differentiation of the regenerate is governed by the stump of the amputated organ, and not by influences originating in the body as a whole. Forelimbs have been grafted into the place of a hind-limb, and amputated after some time. In such cases a fore-limb was regenerated on the stump. If the limb had been rotated during transplantation (e.g., implanted upside down), the orientation of the regenerate agreed with that of the stump, and not with that of the body (Weiss, 1924). We have already seen (p. 159) what will happen if a limb or tail is extirpated completely and no stump left at all: no regeneration takes place in this case. This led Guyénot (1927) to the formulation of the concept of *"territoire de régénération"*. For each part of the body, a certain region can be indicated within which regeneration of this part can take place; if this region is removed completely, no regeneration occurs. Within the region, the formation of the part in question can even be induced by local stimulation, e.g. by an outgrowing or regenerating nerve. If a nerve of the hind-limb (nervus ischiadicus) in amphibians is diverted under the skin in the vicinity of the fore-limb, a supernumerary fore-limb is formed there. If diverted under the skin of the tail, it causes the formation of an extra tail (Guyénot, 1928).

The fact that the organs developing in the regenerate are in direct connection with the remaining organs of the stump might lead to the assumption that the former originate simply as outgrowths of the latter. In other words, each tissue that is exposed in the wound would grow out into the regenerate, and in this way the replacement of the missing part would be brought about by the co-operation of the organs. The following experiments disprove this view. Weiss (1925) extirpated the

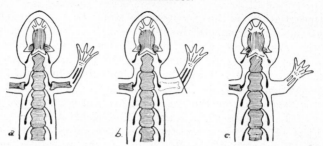

Fig. 62. Diagram of fore-limb regeneration in **Triton**. (*a*) normal skeleton of the fore-limb; (*b*) extirpation of the humerus, followed by the amputation of lower-arm and hand; (*c*) the skeleton of the regenerated lower-arm and hand is normal, but the extirpated humerus is not regenerated. After Przibram.

humerus from the fore-limb of a newt, and a little later amputated this limb in the upper-arm region. No skeleton was present in the wound area, therefore. Nevertheless a normal forearm and hand skeleton were formed in the regenerate, whereas in the upper arm the extirpated humerus was not regenerated (Fig. 62). Therefore, the skeleton in the regenerate is certainly not formed by the outgrowing skeleton of the wound; it must have differentiated on the spot from the regeneration material. Weiss has made similar experiments on skin. The same point was illustrated even more clearly in an experiment by Umanski (1938). He irradiated a limb of a black axolotl with X-rays, thereby destroying the regenerative capacity of its tissues. The skin of this limb was then replaced by that of a normal specimen of the white axolotl. The limb was thereupon amputated, and regeneration took place. The lack of pigmentation in the regenerate seems to show that it originated from the grafted skin which had retained its power of proliferation. Similar experiments were made by Trampusch (1951). Instead of skin, he grafted in other cases skeleton or muscle tissue into the irradiated limb. It appeared that in these cases, too, regeneration occurred.

The following experiments supply some information on the degree of determination in the regeneration buds. Young buds, which have just become visible as half-spherical outgrowths are resorbed if transplanted into the flank of another animal of the same age. Older buds, on the other hand, go on developing, and produce approximately the same parts as they would have formed if they had not been transplanted. A young regeneration bud of a fore-limb, grafted onto the stump of an amputated hind-limb, develops into a hind-limb regenerate (Milojevic, 1924), (Fig. 63, d-e). A young tail regeneration bud produces a fore-limb if grafted into the neighbourhood of a fore-limb, though an older tail regeneration bud develops into a tail (Weiss, 1927). Conversely, a young limb regeneration bud forms a tail if transplanted onto the base of a tail (Guyénot, 1927). These experiments show that young regeneration buds are still more or less indifferent, and not yet determined for a definite course of development. After transplantation, they develop in agreement with their new environment, evidently under the influence of this environment.

Older regeneration buds, on the other hand, have already

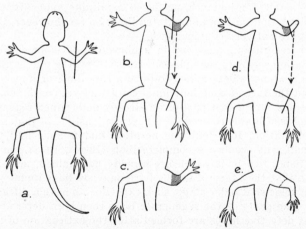

Fig. 63. Transplantation of a regeneration bud. (*a*) amputation of a fore-limb. If the regeneration bud alone is transplanted onto the stump of an amputated hind-limb (*d*), it will form a hind-limb (*e*), but if part of the original stump (hatched) is transplanted as well (*b*), a fore-limb will be formed by the graft (*c*). After Przibram.

been determined for development in a definite direction. After transplantation they produce the same material as they would also have formed if they had not been displaced. The moment of determination at which the buds lose their indifferent character depends on various circumstances; on the average, the indifferent stage of limb buds lasts about two weeks.

Some Russian authors have in later years raised objections to these experiments and their explanation (e.g., Polezajev, Liosner). They pointed out that in many of these experiments the possibility was not excluded that the graft did not go on developing, but that it was repressed and replaced by the stump's own regeneration bud. This would then explain why the properties of the regenerate are in accordance with those of the stump, regardless of the origin of the graft. In reality, early regeneration buds would not be indifferent, but possess a certain degree of labile determination already. In support of this view they described experiments the results of which did not agree with those mentioned above. An early regeneration bud of a tail, for instance, produced a tail-like appendix after transplantation on to an amputated limb. So far, however, their results are not very convincing.

In regeneration, just as in embryonic development, determination seems to take place stepwise. First the regeneration bud as a whole is determined to form a fore-limb, or hind-limb, etc., but the fate of each individual cell is not yet fixed. Schaxel (1922) has shown that part of an already determined regeneration bud is still able to produce an harmoniously built limb after transplantation. If a limb regeneration bud is split into two, each half will produce a complete limb (Swett, 1928). On the other hand, two buds which are already determined, may still fuse so that they form a single, harmoniously built limb.

At a somewhat older stage, such experiments have a different result. The parts of the regenerate have then been determined, so that defective limbs are formed after the extirpation of part of the material, or splitting of the bud.

As regards the problem of how the determination of the various organs within the regenerate takes place, it can be stated on the basis of the experiments already described

(p. 185) that this process does not occur in such a way that the determination of each tissue of the regenerate is governed by the cells of similar tissues in the stump. The subdivision of the regenerate into its organs is a process entirely similar to the autonomisation of the organ-fields in embryonic development. Each part of the regeneration material is as though "imprinted" with a certain determination, depending on its position in the system. The fate of each element is a function of its position. Here again we are dealing with an organisation-field.

If we want to solve the problem of the origin of this field, we must realise that the regenerate replaces the part that has been removed by amputation, i.e. that its development varies in accordance with the position of the cut. Therefore, the field governing the development of the regenerate must also vary from case to case. The simplest explanation seems to be that, e.g. in the limb of the adult animal, the original limb-field is retained intact. After amputation, it is still present in the stump. From there, it extends into the indifferent material of the regeneration bud, and determines the differentiation of this material. The field in the regenerate, therefore, would be the immediate continuation of the field in the stump, and this would explain the complete harmony between the regenerated organs and those in the stump.

Certain experimental results, however, conflict with this explanation. We have seen that an early regeneration cone of a fore-limb, grafted on to the stump of a hind-limb, differentiates into a hind-limb, in accordance with its new environment. If however a thin slice of the *old* tissue of the fore-limb is transplanted along with the regenerate, the latter will develop into a fore-limb (Milojevic, 1924), (Fig. 63). The disc of fore-limb tissue, therefore, prevents the influence of the hind-limb stump from taking effect, and gives rise to a fore-limb field in the regenerate. In this case the field is certainly not the immediate continuation of that of the stump. Weiss (1927) amputated the hand of a fore-limb, halved the lower arm longitudinally, and removed one of the halves. He then covered the lateral wound with skin, so that regeneration could not take place in this region, but only at the distal transverse cut. A complete hand was then formed by the regenerate on the halved lower-arm. Here, again, the field of the regenerate could not be simply the continuation of that of the stump. The opposite experiment can also be made: if two limbs are made to grow together longitudinally with their axes pointing in the same direction, and then amputated, a single regenerate will develop in the common wound (Swett, 1924). Polezajev (1936) went even farther, by completely mincing the internal organs of the stump of an amputated limb. Yet a more or less complete regeneration of the limb was found to take place. If, however, the stump was filled with minced tail tissue, tail-like regenerates were formed. It follows that it is not necessary for the formation of a normal organisation-field in the regenerate that the organs of the stump should be in their normal positions. The field is clearly not an extension of a pre-existing field in the stump, but it is formed *de novo* in the regenerating tissue by an autonomous process. The composition of the field is such that the organs differentiating under its influence fit exactly on to those of the stump. The following experiment proves that the nature of the field is determined only by the level of the cut, and not by the composition of the rest of the stump. Part of a limb, e.g. a forearm, is isolated by means of two transverse cuts, one at the elbow, and one at the wrist. It is then put back on to the stump in an inverted position, so that the elbow is at the distal end. Now at this end, a regenerate is formed consisting of the parts that normally lie distally to this cut, i.e. forearm as well as hand; a new forearm is consequently formed in the regenerate, in spite of the fact that an (inverted) forearm is present already. This must be due to the "elbow" qualities of the wound in which regeneration takes place.

Faber (1960) studied the differentiation of limb regeneration-buds of various age transplanted to the flank. Early buds formed merely structures of distal character (digits), though its material would mainly have developed into proximal parts if left in place. In older buds more proximal skeletal elements were formed, while the capacity to form a qualitatively complete limb is only acquired by an advanced paddle-shaped blastema. Faber concluded from these experiments that the early blastema cells produced by de-differentiation of the stump tissues constitute an autonomous proliferation and organisation centre, which continuously forms new mesenchymal material possessing tendencies to the differentiation of distal structures. This centre remains situated at the tip of the blastema during elongation. In the more proximal regions of the elongating blastema the original distal differentiation tendencies of the material are gradually changed into more proximal ones by influences of the stump, spreading with decrement through the blastema. Therefore the regional organisation of the regenerating limb is the result of an interaction between the apical organisation centre and the basal stump influence.

Experiments by Goss (1956) have shown that, superimposed upon the action of the regeneration field, other influences, of the nature of a tissue-specific induction, also play a part. While the absence of one or more of the bones in the wound area does not preclude the normal replacement of the missing skeletal parts (Fig. 62), on the other hand following the addition of extra bones to the limb and subsequent amputation through the treated region, corresponding supernumerary skeletal elements are produced in the regenerate. The two mechanisms, field action and tissue-specific induction, are regarded as more or less complementary.

Liosner and Woronzowa (e.g., 1937) have made a number of experiments in an attempt to analyse the complex of determining factors that originates in the wound. They transplanted either only musculature or only skeleton from fore-limb into hind-limb, from tail into limb, from upper arm into forearm, and *vice versa*. After this, they amputated the part in question. During regeneration they found that both muscle grafts and skeleton grafts influenced the nature of the regenerate. They concluded that the determination of the regenerate is brought about by the co-operation of the various transected organs in the wound.

Trampusch (1958) made similar experiments, but first suppressed the capacity of regeneration of the host limb or tail by X-irradiation. While implantation of skin or skeleton of a non-irradiated limb into an irradiated one reactivated the regenerative abilities of the latter, so that it regenerated a new limb after amputation, skin of the body and mandibular bone remained without effect, while skin or skeleton from the tail evoked regeneration of tail-like outgrowths. Conversely, transplantation of limb skin into an irradiated tail, which was then amputated through the graft, produced regeneration of a limb. Trampusch concluded that the inability of an irradiated limb or tail to regenerate is due to the extinction of its morphogenetic field; this enables the non-irradiated graft to act as a dominating morphogenetic factor, determining the shape of the regenerate as its organiser.

From similar experiments Trampusch (1951) concluded that the skeleton is important in particular for the longitudinal growth of the regenerating limb, and the muscle tissue for its increase in girth, but that the skin is the most important organiser for the shape of the regenerate.

The latter is confirmed by experiments of Goss (1957) showing that the regulatory capacity of a regenerating limb which had been halved longitudinally is dependent upon the presence of a full complement of skin on the stump.

In tail regeneration, on the other hand, it is especially the spinal cord which plays the main role. The outgrowing spinal cord induces the formation of the new axial skeleton from the blastema, and is responsible for the segmentation of the vertebral column and the tail musculature (Holtzer, 1956).

Summarising the foregoing, we can say that in many respects the phenomena of regeneration show a parallel to those of embryonic development, but that in a number of respects we have found new relationships here. These throw some light from a fresh angle on the problem of the organisation-field. We may expect a deeper insight into the laws governing development from the further analysis of these phenomena.

CHAPTER XII

Some final considerations

How does the spatial multiplicity of the adult organism arise in the course of its development? This is the central problem, underlying the whole of the present discussion. We have seen that the spatial multiplicity as such is not yet present in the fertilised egg. The latter shows only a slight extensive multiplicity, as expressed in its polarity and bilateral symmetry, and probably located mainly in the egg cortex. Development, however, involves a steady increase in the complexity of the structure of the embryo. This begins with the local accumulation of preformed determining substances under the influence of directive activities originating in the primary coordinate system. The cells arising by cleavage are thus endowed from the beginning with different proportions of cytoplasmic substances and with different parts of the original cortex. This primary chemodifferentiation leads to a differential activation of genes in the nuclei, by which different synthetic processes are set going. This, in turn, starts certain topogenetic processes which result in migrations of cell material, and thereby create new topographical relationships which set the scene for new mutual influences, called induction. Finally, the physical and chemical variety of the cells caused by their chemodifferentiation becomes outwardly visible. This marks the beginning of tissue differentiation, which prepares the organs for their definitive functions in the organism. Even after the attainment of the functional stage, development does not come to a stop, but it goes on under continuous interaction of the parts. Last of all, the phenomena of regeneration show that even in the adult animal the parts still interact ceaselessly: removal of a part of the body causes new processes in the remaining part which result in complete or partial restitution of the missing part.

To a very great extent, therefore, development has the character of an *epigenesis*. The spatial structure of the adult organism is not preformed as such in the egg, but it arises *de novo* in the development of each individual. We must not forget, however, that the multiplicity is already present in the egg in another form, viz. as intensive multiplicity, as expressed by the complexity of the substance mixture of the cytoplasm, and as structural multiplicity at a molecular and submicroscopic level. But for this pre-existing multiplicity of the egg cell, its further development would be inconceivable. The spatial multiplicity of the animal, therefore, does not arise from nothing at all, but rather by the transformation of intensive into extensive, and of molecular into macroscopic multiplicity. In other words, the structural plan is potentially present in the egg already, though not, it is true, in its ultimate form; it is actualised during development. Ultimately, all developmental processes can be referred back to the constitution of the egg, from which they follow according to fixed laws. This is the cause of their harmonious interlinking, which is so essential for the normal course of development. The orderly character of development, therefore, is due to the constant composition of the egg, which is given once and for all. The secret of development lies in the composition of the fertilised egg; from it, all the rest follows of necessity.

It can easily be understood that the consideration of the phenomena of development, in which the marvellously perfect structure of the organism seems to arise from nothing, has given rise to questions regarding the nature of this great mystery: Life. The very first investigations in the field of "developmental mechanics" were designed to shed some light on this problem. Roux, the "father of developmental mechanics", himself adhered to the then dominant *"machine theory"* of life, which regarded the living organism as nothing but a well functioning machine. On this view, all biological phenomena would be explicable entirely by means of the laws of physics and chemistry. The egg, too, was regarded as such a complicated machine; its development would be nothing but the setting in motion of a wound-up clockwork. Weismann's theory, which we have discussed above (p. 38), followed this train of thought to its final logical conclusion.

A campaign against these conceptions was started by Driesch. His experiments, in particular those on sea urchins, had convinced him that the machine theory could not explain the phenomena of development. The chief obstacle to this theory was the explanation of the phenomena of regulation, whereby normal embryos are formed in spite of disturbances during development. The machine theory was not able to make this comprehensible. According to Driesch it was not possible to conceive a machine which can be divided into any number of machines of the same structure, or which, after a disturbance of its structure or functioning, returns to the normal situation. Hence he concluded that, apart from the "machine" there was another, non-mechanical factor, which he called *"entelechy"*. It was this factor that, after disturbances, was able to repair the machine and to make it return to the right track. He opposed this theory, a form of *"vitalism"*, to the *"mechanicism"* of Roux and Weismann. Because of the interference of the "vital force", or entelechy, in the mechanical course of the life phenomena, no purely physico-chemical explanation of these phenomena would be possible. This was the basis of Driesch's conviction that he had proved the *"autonomy of life"*.

We have already seen (p. 35) that this "proof" by Driesch is not valid. His conclusion that, apart from the machine, the assumption of a special vital force is necessary to explain regulation, is not the only possible conclusion, nor even the first one that comes to mind. The phenomena of regulation can be explained far more easily by the hypothesis that, at the stage in question, there is no machine at all as yet, i.e. that no system of spatial multiplicity has yet been formed. This disposes of Driesch's arguments, and his vitalism thereby loses its *raison d'être* as an empirically founded theory.

Now does this mean that "mechanicism" is the only correct view on the nature of life, and that, therefore, an explanation in terms of physics and chemistry is the only legitimate one, and gives a complete and satisfactory understanding of the phenomena of life? In this discussion, we have always strongly emphasised the physical and chemical processes in the developing organism. Hence it might be assumed that the author supports this view. Such, however, is not the case; and for that reason we must devote our attention for a moment to the relation between biological and physicochemical systems, and to the conclusion on the nature of life to which this leads. We may start from the considerations developed by H. J. Jordan (1941, and elsewhere).

In Jordan's opinion, the difference between physics and chemistry, on the one hand, and biology, on the other, is a difference in *method* rather than a difference in *subject*. It is not really correct to oppose physics and chemistry, as the sciences of lifeless nature, to biology, as that of living nature. Physics and chemistry are sciences of the whole of nature, both living and lifeless. Their task is the study of the *modes of operation* occurring in nature. Therefore, being analytic and experimental sciences, they attempt as far as possible to isolate the factors acting on one another. They attempt to become independent of the fortuitous circumstances of the experiment by eliminating, one after the other, all those factors which interfere with the analysis of the phenomenon that is being studied. In this way, physics and chemistry arrive at formulations of "laws of nature", which are expressions of the modes of operation encountered. These sciences are not interested in the fact that the circumstances postulated in these "laws" do not "occur" in nature in this form, so that no falling object behaves entirely according to the simple law of gravitation, and no billiard ball is so perfect that it does not show deviations from the basic laws of collision. What is more, they are scarcely, if at all interested in the actual *occurrence,* the "here and now" of the phenomena. Physics studies the properties of electrical discharges, but does not ask where the lightning strikes. Chemistry states that A reacts with B if the two meet, but it is not interested in knowing whether or not the two ever do meet in nature without the mediation of the investigator, and whether this occurs often or rarely.

In biology, the situation is entirely different. Here all interest is focused on the *"here"* and *"now"*. It is not the fact that A reacts with B which is the most important thing in biology, but that this reaction takes place exactly *here*, and at just *this* moment. All phenomena in living organisms have their fixed place and time of occurrence; only by virtue of this fact is the orderly progress of living phenomena possible. Interest centres on this *orderliness* in biology, whereas physics and chemistry consider only isolated modes of action and are blind to their association in the real process.

It follows from these considerations that we must divide the question of the relationship between biology, physics and chemistry into two parts. We must consider, on the one hand, the modes of action occurring in living organisms, as such. But, on the other hand, we must take into account their spatial and temporal connections.

It follows from the above that it is meaningless to ask whether in biology modes of action will at any time be found which for fundamental reasons cannot be reduced to, and explained by, the laws of physics and chemistry. For if the biologist, in the course of his analyses of living organisms, encounters modes of action that are not yet known to, or analysed by, the physicist or chemist, it will be the task of the latter to study these phenomena with their own methods, i.e. by complete isolation, and to trace the laws on which they are founded. This is not just a hypothetical case, as may be seen from the number of times that physics and chemistry have already borrowed their problems from biology. A well known example is the importance of the role played in the development of the theory of electricity by Galvani's observation that exposed frog muscles twitch if they touch two different metal conductors. In a later period, physical chemistry received an important impetus from Pfeffer's observations on osmotic phenomena in plant cells, which led to a closer analysis by physicists and chemists of the laws on which these phenomena are founded. It might be said, in general, that the sciences of "here" and "now" (to which belong, apart from biology, such other branches of science as geology, physical geography, meteorology, and astronomy), supply the material that is studied more closely by physics and chemistry. These latter disciplines unravel this material into its elementary components, in order to derive the natural laws that are valid for the whole of the reality of nature. Once we have understood this, we see that the assumption is unfounded that biology may some time detect modes of action which, for fundamental reasons, do not belong to the domain of physics and chemistry, and which these sciences will be unable to analyse. On the other hand, it follows that biology, wherever it employs causal analysis, will greatly benefit by the use of the concepts coined by physics and chemistry, and of the laws discovered by these disciplines.

However, the matter takes on an entirely different aspect, if we state the problem as follows: Is it possible to give a complete explanation of the phenomena of life on the basis of physics and chemistry? In other words, is biology nothing but a part of physics and chemistry? We have already seen that what matters in the living organism are not only the modes of action as such, but in particular also their localisation in space and time. The modes of action as such are not different from those studied by physics and chemistry; it is their mutual connections, the "here" and "now", in a word, the *orderliness* of the phenomena of life, that constitutes the typical subject of biology. We need not consider here whether or not non-living systems can have an "order", comparable to the orderliness found in living organisms (Köhler's *"physische Gestalten"*). The important fact is that physics and chemistry are powerless to explain this "order" because it is outside their domain. This "order" cannot be analysed further with the causal method; if we trace the phenomena backwards through time, we find that the orderly progress of the phenomena of life always proceeds from the preformed order of an earlier situation. Without this order, life is not conceivable; biology, therefore, will never be able to dispense with it without belying its most essential nature.

If we apply these ideas to animal development, we see that the phenomena as such, taken severally, can be described simply as physical and chemical processes. Movements of material particles, diffusion and osmosis, chemical reactions, colloid-chemical processes — physical and chemical laws govern the course of all these phenomena. But as soon as we focus our attention on the orderly character of the developmental phenomena, the way in which they are harmoniously inter-linked, their well-arranged course in space and time — character-istics which constitute the most essential trait of development, because it is only by their virtue that the result is not a chaotic aggregate, but an *organism* — as soon as we do this, physics and chemistry fail us. We can reduce the orderly course of development only to preformed order in the constitution of its starting point, the egg, in which the future embryo is contained, not, it is true, in an actually spatial form, but potentially, non-spatially, as an intensive multiplicity.

If now we ask whence this order of the egg originated, we can go still farther back in time, and trace the developmental processes which gave rise to the egg, but always again we encounter the order, given once and for all, which is the characteristic of all life. In the phenomena of life, all order is a consequence of previous order.

Until quite recently, biologists were rather powerless against this organismic order. They could draw the attention to the orderly and highly efficient structure and functioning of living things, but did hardly proceed beyond this point. The concepts relating to this final aspect of life processes, like fitness, whole-ness, and so on, were lacking in precision and did not lend themselves to an exact quantitative treatment. In his efforts to give a causal analysis of life processes, the scientist easily lost sight of this aspect; the wholeness of living organisms was analysed to pieces, the fitness dissolved itself into physical and chemical actions between elementary components. There-fore, these problems, which belong to the most fundamental ones in biology, as a rule did not get the attention that they deserved.

Recently, however, important new tools for the study of these relationships have become available to the biologist by the development of *cybernetics* and *information theory*. The concept of *feedback* as used in cybernetics forms a convenient model for all kinds of regulatory processes which are so common in the living organism, and, as we have seen, also play a great part in embryonic development. On the other hand, the concept of *information* may be used as a mathematical measure of the degree of orderliness of a system, and thereby enables a more exact treatment of the final aspect of life processes.

Applying this to our problem, we can say that the relationship between the generations of living organisms, connected by sexual reproduction, in which the ordered structure of the parents is repeated in their offspring, can be treated as an example of the transmission of information. Information is transmitted from parents to offspring through the sex cells and the fertilised egg cell produced by their union. The information (order) contained in the structure of the adult animal, as far as it is hereditary, must be present in form of a *code* in the fertilised egg from which it develops. It is encoded during the formation of the egg cell, stored in its structure, and decoded during development, which means a translation of the code into the ordered structure of the adult.

We have seen above (p. 94) that part of the developmental information stored in the egg cell is probably contained in the molecular structure of the DNA of the chromosomes. Other parts of this information are probably carried by the cytoplasm and the cortex, so that the developmental code is subdivided into three complementary sets of instructions, which are assigned to the three main components of the egg cell (Raven, 1961).

It is probable that the problem of orderliness in development can be treated by means of information theory in a more exact and satisfactory way. Further progress in this field must be awaited, however.

Glossary

The following list contains the explanations of a number of biological technical terms used in this book.

Acrosome: body at the anterior end of a spermatozoön.

Amoeboid movement: movement by means of pseudopodia, similar to that of an Amoeba.

Amphiaster: double star, consisting of two asters connected by a mitotic spindle. Cf. monaster.

Amphimixis: fusion of two gametes (in particular: fusion of their nuclei), formed by two different individuals.

Animal pole: pole of the egg at which the polar bodies are formed. Opp.: vegetative pole.

Antifertilizin: fertilisation substance produced by the male germ cells. Cf. fertilizin.

Antigen: a substance which can give rise to the production of antibodies.

Assimilative induction: influence which causes other cells to develop in the same way as those of the inductor.

Aster: star-shaped structure in the cytoplasm, formed by local gelation.

Atrophy: reduction in size and functional activity. Cf. hypertrophy.

Autonomisation: process leading to independence of the parts of a whole.

Bilateral symmetry: case of symmetry in which there is only one symmetry plane, i.e. in which the body can be divided into symmetrical halves in only one way.

Blastema: organ primordium consisting of undifferentiated cells.

Blastomere: cell formed by the cleavage of the egg.

Blastula: vesicular stage of cleavage, with an internal cavity.

Caudal: pertaining to or directed towards the tail. Opp.: cranial, rostral.

Centrifuge: rapidly rotating apparatus for subjecting bodies to strong centrifugal forces.

Centriole: corpuscle in a cytocentre, supposed to play a role in cell division.

Centrosome: corpuscle forming the centre of an aster, or a pole of a mitotic spindle. Cf. cytocentre.

Chemodifferentiation: division of a whole into parts differing in their (physical and) chemical properties.

Chimera: organism consisting of tissues belonging to different species.

Chromatin: substance with a high affinity for stains, localised in the nucleus.

Chromosome: corpuscle containing the chromatin during nuclear division.

Cilia: hair-like, motile cytoplasmic processes.

Cleavage: the division of the egg into a number of cells by a process of cellular division, which is not accompanied by cell growth.

Code: a system of signs carrying information.

Competence: reactivity of a cell group with respect to an induction.

Cortex: outer layer.

Cortical: pertaining to the cortex.

Cranial: pertaining to or directed towards the head. Opp.: caudal.

Cyclopia: abnormality, in which only one median eye is present.

Cytaster: aster formed at an arbitrary place in the cytoplasm.

Cytocentre: corpuscle or region of a cell, around which an aster may develop. Cf. centrosome.

Cytology: the science of cells.

Cytoplasm: protoplasm of the cell, as opposed to that of the nucleus (cf. nucleoplasm).

Decoding: translation of a coded message into another language.

De-differentiation: reduction in the differentiation that has already taken place.

Deficiency: loss of a part of a chromosome.

Depolarisation: loss or reduction of polarity.

Determinants: material particles, regarded as the carriers of inheritable properties. Cf. genes.

Determination: the fixation of the course of future development in a part of the germ.

Differentiation: the division of a whole into recognisable parts; the appearance of differences among originally identical parts.

Diploid: containing a double set of chromosomes. This is the usual condition in the majority of the cells of the body. Cf. haploid, polyploid.

Dispermy: Fertilisation by two sperms. Cf. monospermy, polyspermy.

Double-monster: organism formed by two individuals grown together.

Ectoderm: outer germ-layer, the outer cell-layer of the gastrula. Cf. mesoderm, endoderm.

Ectoplasm: outer layer of the cytoplasm; also, protoplasm which will later be located in the ectoderm cells. Opp.: endoplasm.

Embryogenesis: the development of the embryo from the egg.

Encoding: translation of a message into a code.

Endoderm: inner germ-layer, inner cell-layer of the gastrula. Cf. ectoderm, mesoderm.

Endoplasm: the central part of the cytoplasm, or: the protoplasm that will later be located in the endoderm cells. Opp.: ectoplasm.

Endoplasmic reticulum: delicate cytoplasmic structure, only visible with the electron microscope.

Endothelium: the tissue lining cavities in the body, such as blood and lymph vessels.

Entelechy: a hypothetical, purposive vital force.

Enzyme: organic substance with the properties of a catalyst; = ferment.

Epigenesis: the appearance of new structures in the course of development.

Epithelium: tissue, consisting of a sheet of closely packed cells, covering an internal or external surface of the body. Cf. endothelium.

Equatorial plane: plane that bisects at right angles the line connecting the poles.

Equipotential system: cell group, all parts of which have the same potency.

Ergastoplasm: part of the cytoplasm having distinct staining properties, which may play a part in the synthesis of certain cell products.

Evocator: substance produced by an inductor, and influencing the determination of other cells.

Exogastrulation: abnormal gastrulation, in the course of which the archenteron evaginates, instead of invaginating.

Explantation: the culturing of pieces of tissue outside the body. Cf. implantation, transplantation.

Extensive multiplicity: spatial multiplicity: a complexity of structure which is recognisably different in form or nature in different regions of an organism or cell. Opp.: intensive multiplicity.

False hybrid: hybrid formed by the fusion of an egg with a sperm of another species, in which the nucleus of the latter does not take part in development.

Ferment: = enzyme.

Fertilizin: fertilisation substance secreted by the female germ cells. Cf. antifertilizin.

Fibroblast: young connective tissue cell.

Fibula: splint-bone.

Follicle: multicellular covering of the egg in the ovary.

Frontal: at right angles to the median plane, separating dorsal and ventral halves.

Gamete: germ cell.

Gastrula: developmental stage, consisting of 2 (or 3) germ-layers.

Gastrulation: development of a gastrula from a blastula.

Gene: unit carrier of inheritable properties. Cf. determinants.

Genotype: the inherited constitution of the individual.

Germ-band: thickening of one or several germ-layers, from which certain organs will develop.

Germinal vesicle: the nucleus of the oöcyte.

Germ-layer: one of the cell-layers of the gastrula, each of which will later develop into definite organs.

Golgi apparatus: cell organelle characterised by its great affinity for osmium and other metals.

Gradient: change of a scalar quantity from point to point.

Haploid: containing a single set of chromosomes. This is the normal condition in ripe germ cells. Cf. diploid, polyploid.

Heterogeneous fertilisation: fertilisation by a sperm of another, usually distantly related, species.

Heteroplastic transplantation: transplantation into an individual of another, closely related species. Cf. xenoplastic transplantation.

Heterospermic merogony: development of a non-nucleated egg fragment after fertilisation by a sperm from another species. Opp.: homospermic merogony.

Heterozygote: organism developing from a zygote formed by the fusion of gametes that were dissimilar as regards the gene under consideration. Opp.: homozygote.

Histochemistry: the determination of the chemical composition of cells and tissues.

Histogenesis: the origin of tissues.

Homospermic merogony: development of a non-nucleated egg fragment after fertilisation by a sperm of the same species. Opp.: heterospermic merogony.

Homozygote: organism arising from a zygote formed by the fusion of gametes that were identical as regards the gene under consideration. Opp.: heterozygote.

Hormones: substances secreted in definite parts of the body, under the influence of which certain processes take place in other parts of the body to which they are carried by the blood.

Hybrid: individual arising from a cross between two races or species.

Hypertrophy: excessive growth by the increase in size of tissue elements.

Implantation: the grafting of a cell group or organ into an abnormal place, in particular into one of the body cavities. Cf. explantation, transplantation.

Induction: the influence exerted by an inductor.

Inductor: cell group which causes other cells to develop into a certain direction.

Information: the order contained in a message.

Intensive multiplicity: A complexity of structure which is not based on recognisable regional differences in form or nature within the organism or cell. Opp.: extensive multiplicity.

Lysin: enzyme having a dissolving action on certain substances or membranes.

Lysosomes: vesicular bodies in the cytoplasm containing hydrolytic enzymes.

Malpighian tubes: excretory organs of insects, opening into the hind gut.

Matrocline: in which the maternal characteristics dominate over the paternal ones. Opp.: patrocline.

Median plane: plane of symmetry of a bilaterally symmetrical organism.

Melanophore: black pigment cell.

Mendelian inheritance: inheritance according to the rules first established by G. Mendel.

Meridian plane: plane containing both poles.

Merogony: development of a fertilised fragment of an egg in which no female pronucleus is present.

Mesenchyme: embryonic connective tissue.

Mesoderm: middle germ-layer. middle cell-layer of the gastrula. Cf. ectoderm, endoderm.

Mesoplasm: cytoplasm that will later become located in the mesoderm cells. Cf. ectoplasm, endoplasm.

Microdissection: operations on cells under the microscope.

Micromeres: small cleavage cells.

Microsomes: very small granules in the cytoplasm.

Mitochondria: larger, rod-shaped or granular corpuscles in the cell cytoplasm.

Mitosis: indirect nuclear division.

Modulation: change in the external appearance of cells, caused by external circumstances, and not accompanied by a simultaneous change in potencies.

Monaster: simple radiation in the cytoplasm, in the centre of which the chromosomes are located. Cf. amphiaster.

Monospermy: fertilisation by a single sperm. Cf. dispermy, polyspermy.

Morphallaxis: restitution by the transformation and shifting of material, not by growth. Cf. reorganisation.

Morphogenesis: the origin of form.

Morphology: the science of form.

Mutant: individual or race arising by mutation.

Mutation: spontaneous change in an inheritable property.

Neoblast: undifferentiated cell in the body of an adult animal, which plays a role in regeneration.

Neo-epigenesis: the hypothesis that the embryo develops from a structureless egg. Cf. Neo-evolutio.

Neo-evolutio: the hypothesis that the egg already contains an invisible spatial structure, from which the embryo will later develop. Cf. Neo-epigenesis.

Neural: pertaining to the nervous system.

Neurula: developmental stage, subsequent to the gastrula stage, in which the primordium of the central nervous system (the neural plate) has developed from the ectoderm.

Neurulation: the development of a neurula from a gastrula.

Notochord: axial rod, in the body of Tunicates and Vertebrates.

Nucleoplasm: non-chromatic content of the nucleus, nuclear sap.

Odontoblasts: cells of a tooth germ producing dentine.

Omentum: a fold of the peritoneum.

Omnipotent: capable of development into any organ or tissue.

Oöcyte: unripe egg, before the completion of the maturation divisions.

Oögenesis: the growth of the egg cells in the ovary.

Oögonium: mother cell of the egg.

Oöplasmic segregation: partial separation of the substance mixture of the egg cytoplasm during early development.

Organelle: structural component of a cell with a special function.

Organisation: arrangement of the parts of a whole into an orderly system of organs.

Organisation centre: part of the germ, which plays a leading role in the organisation of the embryo.

Organiser: cell group which incites tissues to form an organised system of organs.

Organogenesis: the development of organs.

Ovary: female reproductive gland. Cf. testis.

Oxidase: enzyme promoting oxidation by oxygen. Cf. peroxidase.

Parthenogenesis: reproduction by means of gametes which develop without fertilisation.

Patrocline: in which the paternal characteristics dominate over the maternal ones. Opp.: matrocline.

Peristome: mouth area.

Perivitelline space: the interspace between the egg and the fertilisation membrane.

Peroxidase: enzyme which liberates oxygen from peroxides. Cf. oxidase.

pH: degree of acidity.

Pharynx: part of the intestine, immediately caudal to the mouth cavity.

Physiology: the causal science of the phenomena of life.

Placode: thickening of the ectoderm.

Plasmalemma: extremely thin membrane at the surface of cells.

Pluteus: larva of a sea urchin or brittle star, with a calcareous skeleton consisting of radially protruding rods.

Polar bodies: small cells expelled from the egg in the course of its maturation divisions.

Polarity: condition of the body in which an axis connecting two different poles can be distinguished.

Polyploid: containing more than two sets of chromosomes. Cf. triploid.

Polyspermy: the penetration of several sperms into a single egg. Cf. dispermy, monospermy.

Postpharyngeal: caudal to the pharynx. Opp.: prepharyngeal.

Posttrochal: basal to the band of cilia in the trochophore larva. Opp.: pretrochal.

Potency: power of development.

Predetermination: first phase in the process of determination, during which the fixation of the fate of the cells is still labile.

Preformation: presence in the egg of the structure that will become unfolded during the development of the embryo.

Prepharyngeal: rostral to the pharynx. Opp.: postpharyngeal.

Pretrochal: situated above the band of cilia in the trochophore larva. Opp.: posttrochal.

Pronucleus: nucleus of a gamete.

Proteolytic: causing the breakdown of proteins.

Protoplasm: the substance of which living cells consist.

Pseudopodia: motile protoplasmic processes on the surface of cells.

Reaction system: cell group reacting to an induction with a definite developmental process.

Reductional division: nuclear division in the course of which the chromosome number is halved.

Regenerate: cell group formed by regeneration for the restitution of lost organs or parts of the body.

Regeneration: the reformation or restitution of lost organs of parts of the body.

Regulation: in biology, this term is applied in particular to processes that lead to a return to the normal situation after a disturbance.

Reorganisation: restitution by transformation or shifting of material. Cf. Morphallaxis.

Ribosomes: small granules in the cytoplasm, about 150 Å in diameter, and very rich in ribonucleic acid.

Rostral: pertaining to the front end.

Scalar quantity: quantity to which a numerical value can be attributed, but which is not directional.

Somatoblast: cell from which an important part of the body of the embryo will arise.

Sperm aster: aster in the egg cytoplasm arising from a sperm that has penetrated into the egg.

Sperm (atozoön): male germ cell.

Stomodaeum: mouth cavity formed by the ectoderm.

Syncytium: multi-nuclear mass of protoplasm.

Synkarion: zygote nucleus, arising from the fusion of a male and a female pronucleus.

Teloblast: end-cell of a row of cells.

Testis: male reproductive gland. Cf. ovary.

Tetraster: mitotic spindle with four poles.

Tibia: shin-bone.

Topogenesis: the development of the form of the body through the shifting of material during development.

Transplantation: grafting of an organ or cell group into another place. Cf. explantation, implantation.

Triploid: containing three sets of chromosomes. Cf. diploid, haploid.

Trochophore: larval stage, occurring in worms, etc., with crowns of long cilia by means of which it moves.

Trophic: pertaining to the nutrition.

Ultracentrifuge: centrifuge working at very high speeds.

Vacuole: cavity in the cytoplasm, filled with a liquid.

Vegetative pole: one of the poles of the egg. Opp.: animal pole.

Vegetative reproduction: asexual reproduction.

Vitalism: theory according to which the phenomena of life cannot be described entirely in terms of the laws of physics and chemistry.

Vitamins: essential food substances, very small quantities of which are sufficient.

Vitro, in — : "in glass", term applied to the culturing of tissues outside the body.

X-chromosome: sex chromosome, playing a role in the inheritance of sex.

Xenoplastic transplantation: transplantation into an individual of another, very distant, species. Cf. heteroplastic transplantation.

Zygote: fertilised egg, cell formed by the fusion of two gametes.

Zygote nucleus: the nucleus formed by the fusion of the male and female pronuclei.

References

Extensive lists of literature may be found in the works of Dalcq (1941), Spemann (1936), Weiss (1939), Brachet (1944 and 1952), Lehmann (1945), Willier, Weiss and Hamburger (1955) and Waddington (1962). Therefore the following list contains in general only those of the papers quoted in this book, which are not mentioned by or appeared after the publication of these works, to which the reader is referred for all other references.

1955 ALLEN, R. D. and B. HAGSTRÖM: Interruption of the cortical reaction by heat. *Exp. Cell Res.* **9**, 157 (1955).

1948 ANCEL, P. and P. VINTEMBERGER: Recherches sur le déterminisme de la symétrie bilatérale dans l'oeuf des Amphibiens. *Bull. Biol. France Belg.*, Suppl. **31** (1948).

1946 ASPEREN, K. VAN: Pharynx regeneration in postpharyngeal fragments of *Polycelis nigra* (Ehrbg.). *Proc. Kon. Ned. Akad. v. Wetensch., Amsterdam* **49**, 1083 (1946).

1947 AVEL, M.: Les facteurs de la régénération chez les Annélides. *Rev. Suisse Zool.* **54**, 219 (1947).

1948 BALINSKY, B. I.: Korrelationen in der Entwicklung der Mund- und Kiemenregion und des Darmkanals bei Amphibien. *Roux' Archiv* **143**, 365 (1948).

1955 —— Histogenetic and organogenetic processes in the development of specific characters in some South African tadpoles. *J. Embr. exp. Morph.* **3**, 93 (1955).

1957a —— On the factors determining the size of the lens rudiment in amphibian embryos. *J. exp. Zool.* **135**, 255 (1957).

1957b —— New experiments on the mode of action of the limb inductor. *J. exp. Zool.* **134**, 239 (1957).

1947 BALTZER, F.: Weitere Beobachtungen an merogonischen Bastarden der schwarzen und weissen Axolotlrasse. *Rev. Suisse Zool.* **54**, 260 (1947).

1963 BEERMANN, W.: Cytologische Aspekte der Informationsübertragung von den Chromosomen in das Cytoplasma. In: *Induktion und Morphogenese. 13. Colloquium der Gesellschaft für physiol. Chemie.* Berlin-Göttingen-Heidelberg (1963).

1944 BRACHET, J. *Embryologie chimique,* Paris — Liège, 1944.

1952 —— Le rôle des acides nucléiques dans la vie de la cellule et de l'embryon. *Actualités biochim.,* Liège — Paris, 1952.

1957 BRIGGS, R. and T. J. KING: Changes in the nuclei of differentiating endoderm cells as revealed by nuclear transplantation. *J. Morph.* **100**, 269 (1957).

1955 BRØNDSTED, H. V.: Planarian regeneration. *Biol. Rev.* **30**, 65 (1955).

1949 BUTLER, E. G. and O. E. SCHOTTÉ: Effects of delayed denervation on regenerative activity in limbs of Urodele larvae. *J. exp. Zool.* **112**, 361 (1949).

1961 COLWIN, L. H. and A. L. COLWIN: Changes in the spermatozoon during fertilization in *Hydroides hexagonus* (Annelida). I. Passage of the acrosomal region through the vitelline membrane. II. Incorporation with the egg. *J. bioph. bioch. Cytol.* **10**, 231, 255 (1961).

1949 COSTELLO, D. P.: The relations of the plasma membrane, vitelline membrane, and jelly in the egg of *Nereis limbata. J. gen. Physiol.* **32**, 351 (1949).

1963 CURTIS, A. S. G.: The cell cortex. *Endeavour* **22**, 134 (1963).

1941 DALCQ, A.: *L'oeuf et son dynamisme organisateur.* Paris 1941.

1947 —— The concept of physiological competition (Spiegelman) and the interpretation of vertebrate morphogenesis. *Proc. 6th Intern. Congr. Exp. Cytol.* 1947.

1946 DALTON, H. C.: The role of nucleus and cytoplasm in development of pigment patterns in *Triturus. J. exp. Zool.* **103**, 169 (1946).

1949 DUBOIS, F.: Contribution à l'étude de la migration des cellules de régénération chez les planaires dulcicoles. *Bull. Biol. France Belg.* **83**, 213 (1949).

1942 EPHRUSSI, B.: Chemistry of "eye color hormones" of *Drosophila. Quart. Rev. Biol.* **17**, 327 (1942).

1954 EYAL-GILADI, H.: Dynamic aspects of neural induction in Amphibia. *Arch. Biol.* **65**, 179 (1954).

1960 FABER, J.: An experimental analysis of regional organization in the regenerating fore limb of the axolotl (*Ambystoma mexicanum*). *Arch. Biol.* **71**, 1 (1960).

1956 GOSS, R. J.: The relation of bone to the histogenesis of cartilage in regenerating forelimbs and tails of adult *Triturus viridescens. J. Morph.* **98**, 89 (1956).

1957 —— The relation of skin to defect regulation in regenerating half limbs. *J. Morph.* **100**, 547 (1957).

1963 GUSTAFSON, T. and L. WOLPERT: The cellular basis of morphogenesis and sea urchin development. *Int. Rev. of Cytol.* **15**, 139 (1963).

1956 HAGGIS, A. J.: Analysis of the determination of the olfactory placode in *Amblystoma punctatum. J. Embr. exp. Morph.* **4**, 120 (1956).

1959 HAMPÉ, A.: Contribution à l'étude du développement et de la régulation des déficiences et des excédents dans la patte de l'embryon de poulet. *Arch. Anat. micr. Morph. exp.* **48**, 345 (1959).

1957 HENZEN, W.: Transplantationen zur entwicklungsphysiologischen Analyse der larvalen Mundorgane bei *Bombinator* und *Triton. Roux' Arch.* 149, 387 (1957).

1943-44 HOLTFRETER, J.: A study of the mechanics of gastrulation. I. *J. exp. Zool.* 94, 261 (1943). — II. *J. exp. Zool.* 95, 171 (1944).

1947 —— Neural induction in explants which have passed through a sublethal cytolysis. *J. exp. Zool.* 106, 197 (1947).

1956 HOLTZER, S. W.: The inductive activity of the spinal cord in urodele tail regeneration. *J. Morph.* 99, 1 (1956).

1941 JORDAN, H. J.: Die theoretischen Grundlagen der Tierphysiologie. *Biblioth. Biotheor.* 1, Leiden, 1941.

1960 KIENY, M.: Rôle inducteur du mésoderme dans la différenciation précoce du bourgeon de membre chez l'embryon de poulet. *J. Embr. exp. Morph.* 8, 457 (1960).

1945 LEHMANN, F. E.: *Einführung in die physiologische Embryologie.* Basel 1945.

1948 —— Zur Entwicklungsphysiologie der Polplasmen des Eies von *Tubifex. Rev. Suisse Zool.* 55, 1 (1948).

1956 LENDER, T.: Recherches expérimentales sur la nature et les propriétés de l'inducteur de la régénération des yeux de la planaire *Polycelis nigra. J. Embr. exp. Morph.* 4, 196 (1956).

1952 LOMBARD, G. L.: An experimental investigation on the action of lithium on amphibian development. Thesis, Utrecht 1952.

1961 LOPASHOV, G. V.: The development of the vertebrate eye in relation to the problem of embryonic induction. *Ann. Zool. Soc. Zool. Bot. Fenn. "Vanamo",* 22, no. 6 (1961).

1949 MOMENT, G. B.: On the relation between growth in length, the formation of new segments, and electric potential in an earthworm. *J. exp. Zool.* 112, 1 (1949).

1947 NIEUWKOOP, P. D.: Investigations on the regional determination of the central nervous system. *J. exp. Biol.* 24, 145 (1947).

1952 —— and others: Activation and organization of the central nervous system in amphibians. *J. exp. Zool.* 120, 1 (1952).

1961 PASTEELS, J.: La réaction corticale de fécondation ou d'activation. *Bull. Soc. Zool. France,* 86, 600 (1961).

1939 PLAGGE, E.: Genabhängige Wirkstoffe bei Tieren. *Ergebn. d. Biol.* 17, 105 (1939).

1945 POULSON, D. F.: Chromosomal control of embryogenesis in *Drosophila. Amer. Natural.* 79, 340 (1945).

1938 RAVEN, CHR. P.: Ueber die Potenz von Gastrulaektoderm nach 24-stündigem Verweilen im äusseren Blatt der dorsalen Urmundlippe. *Roux' Archiv* 137, 661 (1938).

1958 RAVEN, CHR. P.: *Morphogenesis. The analysis of Molluscan development*. London-New York-Paris-Los Angeles, 1958.

1961 ——— *Oogenesis. The storage of developmental information*. Oxford-London-New York-Paris, 1961.

1963 ——— The nature and origin of the cortical morphogenetic field in *Limnaea. Developm. Biol.* **7**, 130 (1963).

1948 RAVEN, CHR. P. and J. C. A. MIGHORST: On the influence of a posterior wound surface on anterior regeneration in *Euplanaria lugubris* (Hesse). *Proc. Kon. Ned. Akad. v. Wetensch., Amsterdam,* **51**, 434 (1948).

1949 REVERBERI, G. and A. MINGANTI: Ulteriori richerche sulla formazione del cervello, degli organi di senso e dei palpi nelle Ascidie. *Riv. di Biol.* **41**, 125 (1949).

1957 ROSE, S. M.: Cellular interaction during differentiation. *Biol. Rev.* **32**, 351 (1957).

1954 ROTHSCHILD, LORD: Polyspermy. *Quart. Rev. Biol.* **29**, 332 (1954).

1952 RUNNSTRÖM, J.: The problems of fertilization as elucidated by work on sea urchins. *The Harvey Lectures,* **46**, 116 (1952).

1955 SALA, M.: Distribution of activating and transforming influences in the archenteron roof during the induction of the nervous system in amphibians. I. Distribution in cranio-caudal direction. *Proc. Kon. Nederl. Akad. Wetensch., Amsterdam,* C **58**, 635 (1955).

1953 SENGEL, PH.: Sur l'induction d'une zone pharyngienne chez la planaire d'eau douce *Dugesia lugubris* O. Schm. *Arch. d'Anat. micr.* **42**, 57 (1953).

1954 SINGER, M.: Induction of regeneration of the forelimb of the postmetamorphic frog by augmentation of the nerve supply. *J. exp. Zool.* **126**, 419 (1954).

1936 SPEMANN, H.: *Experimentelle Beiträge zu einer Theorie der Entwicklung*. Berlin, 1936.

1945 SPIEGELMAN, S.: Physiological competition as a regulatory mechanism in morphogenesis. *Quart. Rev. Biol.* **20**, 121 (1945).

1948 SPOFFORD, W. R.: Observations on the posterior part of the neural plate in *Amblystoma*. II. The inductive effect of the intact posterior part of the chorda-mesodermal axis on competent prospective ectoderm. *J. exp. Zool.* **107**, 123 (1948).

1960 TARDENT, P. E.: Principles governing the process of regeneration in hydroids. *18th Growth Symposium, Soc. f. Study of Developm. a. Growth,* 21 (1960).

1956 THORNTON, C. S.: The relation of epidermal innervation to the regeneration of limb deplants in *Amblystoma* larvae. *J. exp. Zool.* **133**, 281 (1956).

1955 TOIVONEN, S. and L. SAXÉN: Ueber die Induktion des Neuralrohrs bei *Trituruskeimen* als simultane Leistung des Leber- und Knochenmarkgewebes vom Meerschweinchen. *Ann. Acad. Sci. Fennicae* (A IV) **30** (1955).

1955 TOWNES, PH. L. and J. HOLTFRETER: Directed movements and selective adhesion of embryonic amphibian cells. *J. exp. Zool.* **128**, 53 (1955).

1958 TRAMPUSCH, H. A. L.: The action of X-rays on the morphogenetic field. *Proc. Kon. Ned. Akad. Wetensch., Amsterdam,* C **61**, 417, 530 (1958).

1945 TWITTY, V. C.: The developmental analysis of specific pigment patterns. *J. exp. Zool.* **100**, 141 (1945).

1948 TYLER, A.: Fertilization and immunity. *Physiol. Rev.* **28**, 180 (1948).

1954 UBISCH, L. VON: Ueber Seeigelmerogone. *Pubbl. Staz. Zool. Napoli* **25**, 246 (1954).

1962 WADDINGTON, C. H.: *New patterns in genetics and development.* New York-London, 1962.

1955 WAGNER, G.: Chimaerische Zahnanlagen aus *Triton*-Schmelzorgan und *Bombinator*-Papille. *J. Embr. exp. Morph.* **3**, 160 (1955).

1939 WEISS, P.: *Principles of development.* New York, 1939.

1950 —— Perspectives in the field of morphogenesis. *Quart. Rev. Biol.* **25**, 177 (1950).

1955 —— Specificity in growth control. In: *Biological Specificity and Growth.* Princeton, 1955.

1955 WILLIER, B. H., P. A. WEISS and V. HAMBURGER: *Analysis of development.* Philadelphia and London, 1955.

1962 WOERDEMAN, M. W.: Eye-lens development and some of its problems. *Proc. Kon. Ned. Akad. v. Wetensch., Amsterdam,* C **65**, 145 (1962).

1961 WOLFF, E.: Migrations et contacts cellulaires dans la régénération. *Exp. Cell Res.,* Suppl. 8, 246 (1961).

1955 YAMADA, T. and K. TAKATA: An analysis of spino-caudal induction by the guinea pig kidney in the isolated ectoderm of the *Triturus* gastrula. *J. exp. Zool.* **128**, 291 (1955).

1950 YNTEMA, C. L.: An analysis of induction of the ear from foreign ectoderm in the salamander embryo. *J. exp. Zool.* **113**, 211 (1950).

1956 ZWILLING, E.: Interaction between limb bud ectoderm and mesoderm in the chick embryo. *J. exp. Zool.* **132**, 157 (1956).

Author Index

Subject Index

219